Essays in Game Theory

Nimrod Megiddo
Editor

Essays in Game Theory

In Honor of Michael Maschler

With 27 Illustrations

Springer-Verlag
New York Berlin Heidelberg London Paris
Tokyo Hong Kong Barcelona Budapest

Nimrod Megiddo
IBM Almaden Research Center
650 Harry Road
San Jose, CA 95120-6099
USA

Library of Congress Cataloging-in-Publication Data
Megiddo, Nimrod.
Essays in game theory : in honor of Michael Maschler / Nimrod Megiddo.
p. cm.
Includes bibliographical references.
ISBN 0-387-94224-6
1. Game theory. 2. Maschler, Michael, 1927- . I. Maschler, Michael, 1927- . II. Title.
QA269.M44 1995
519.3—dc20 93-48255

Printed on acid-free paper.

Production managed by Bill Imbornoni; manufacturing supervised by Vincent Scelta.
Camera-ready copy provided by the editor.
Printed and bound by Braun-Brumfield, Inc., Ann Arbor, MI.
Printed in the United States of America.

9 8 7 6 5 4 3 2 1

ISBN 0-387-94224-6 Springer-Verlag New York Berlin Heidelberg
ISBN 3-540-94224-6 Springer-Verlag Berlin Heidelberg New York

Preface

This volume comprises a collection of essays dedicated to Michael Maschler on the occasion of his 65th birthday. The articles were presented to Michael at the International Game Theory Conference in Stony Brook in July 1992. I am indebted to the authors for their contributions, the efforts they made, and the care they took while preparing the text for publication. I once again thank the referees for their help in improving the quality of the accepted essays.

To the readers who do not know Michael Maschler, here is a biographical sketch followed by a list of selected publications. Born July 22, 1927 in Jerusalem, he received from the Hebrew University his M.Sc. degree in 1952 and his Ph.D. degree in 1956. At that time Maschler's interest was in mathematical analysis. He switched to game theory in the early 1960's and from the beginning was interested in cooperative games. His first two articles in game theory dealt with powers of coalitions and appeared in 1963.

It is interesting to mention that Maschler was one of the first to conduct experiments in game theory. While he was teaching in a Jerusalem high school in the 1960's he ran experiments on formation of coalitions in games with empty cores. Later he published an article on experiments. Since then, exeperiments in game theory became an important discipline.

Maschler's early collaboration with Robert J. Aumann led to a path-breaking approach to cooperative games with the introduction of the bargaining set. It was the first solution concept guaranteed to be nonempty. Shortly thereafter, Morton Davis and Maschler introduced the kernel, a subset of the bargaining set. Maschler investigated the structure of both the bargaining set and the kernel. He collaborated on this subject with Bezalel Peleg and Lloyd S. Shapley.

During the 1960's Maschler was, among other activities, also involved in research projects supported by the U.S. Air Force Office of Scientific Research, the Office of Naval Research, and by Mathematica at Princeton. The foundations of the theory of repeated games with incomplete information were laid by Aumann, Maschler, and Richard Stearns in Mathematica research reports. One of the first applications of this theory was the problem of gradual disarmament. Maschler's interest in this problem is also reflected in papers on the "Inspector's Game."

In more recent years Maschler worked with Micha Perles on the superadditive solution for bargaining games. He also wrote papers on cost allocation and bankruptcy games and, together with Guillermo Owen, introduced the consistent Shapley value.

Michael Maschler held visiting appointments at top universities such as Cornell, Northwestern, Princeton, Stanford, and UCLA. He was elected Fellow of the Econometric Society, and has been a member of the editorial boards of the leading journals of game theory: *International Journal of Game Theory* and *Games and Economic Behavior*. He was also a member of the editorial boards of *Management Science* and *Journal of Conflict Resolution*.

Maschler has been very active in the education system in Israel. He chaired a curriculum committee for mathematics in elementary schools, was a member of similar committees for high schools, directed projects for teachers, and authored several textbooks.

Nimrod Megiddo

SELECTED PUBLICATIONS OF MICHAEL MASCHLER

[1] "Minimal domains and their Bergman Kernel function," *Pacific J. Math.* **6** (1956) 501–516.

[2] "Classes of minimal and representative domains and their kernel functions," *Pacific J. Math.* **9** (1959) 763–782.

[3] "Bargaining in n-person cooperative games of pairs," in: *Recent Advances in Game Theory*, Princeton University Press, 1962, pp. 161–169.

[4] "Derivatives of the harmonic measures in multiply-connected domains," *Pacific J. Math.* **12** (1962) 637–647.

[5] "An experiment in n-person games," in: *Recent Advances in Game Theory*, Princeton University Press, 1962, pp. 49–56.

[6] "n-person games with only 1, $n-1$, and n-person permissible coalitions," *J. Math. Anal. Appl.* **6** (1963) 230–256.

[7] "The power of a coalition," *Management Science* **10** (1963) 8–29.

[8] "A non-zero-sum game related to a test ban treaty," in: *Applications of Statistical Methodology to Arms Control and Disarmament*, Report of the U.S. Arms Control & Disarmament Agency/ST–3, Washington, D.C., 1963, pp. 237–287.

[9] "The bargaining set for cooperative games" (with R. J. Aumann), in: *Advances in Game Theory*, Princeton University Press, 1964, pp. 443–476.

[10] "Stable payoff configurations for quota games," in: *Advances in Game Theory*, Princeton University Press, 1964, pp. 477–499.

[11] "The kernel of a cooperative game" (with M. Davis), *Naval Research Logistics Quarterly* **12** (1965) 223–259.

[12] "The inequalities which determine the bargaining set $\mathcal{M}_1^{(i)}$," *Israel J. Math.* **4** (1966) 127–134.

[13] "A price leadership method for solving the inspector's non-constant-sum game," *Naval Research Logistics Quarterly* **13** (1966) 11–33.

[14] "A characterization, existence proof and dimension bounds for the kernel of a game" (with B. Peleg), *Pacific J. Math.* **18** (1966) 289–328.

[15] "Game theoretic aspects of gradual disarmament," (with R. J. Aumann), in: *Development of Utility Theory for Arms Control & Disarmament*, Report of the U.S. Arms Control & Disarmament Agency/ST–80, Chapter V, Washington, D.C., (1966), pp. V1–V55.

[16] "The inspector's non-constant-sum game: Its dependence on a system of detectors," *Naval Research Logistics Quarterly* **14** (1967) 275–290.

[17] "The Structure of the kernel of a cooperative game" (with B. Peleg), *SIAM J. App. Math.* **17** (1967) 569–604.

[18] "Repeated games with incomplete information: a survey of recent results" (with R. J. Aumann), in: *Models of Gradual Reduction of Arms*, Report of the U.S. Arms Control & Disarmament Agency/ST–116, Chapter III, Washington D.C., 1967, pp. 287–403.

[19] "Repeated games of incomplete information: the zero-sum extensive case "info (with R. J. Aumann) in: *The Indirect Measurement of Utility*, Report of the U.S. Arms Control & Disarmament Agency/ST–143, Chapter III, Washington, D.C., 1968, pp. 37–116.

[20] "Repeated games of incomplete information: an approach to the non-zero sum case" (with R. J. Aumann and R. E. Stearns), in: *The Indirect Measurement of Utility*, Report of the U. S. Arms Control & Disarmament Agency/ST–143, Chapter IV, Washington, D.C., 1968, pp. 117–216.

[21] "Some thoughts on the minimax principle" (with R. J. Aumann), *Management Science* **18** (1972) pp. P-54–P-63.

[22] "The kernel and bargaining-set for convex games" (with B. Peleg and L. S. Shapley), *International J. Game Theory* **1** (1972) pp. 73–93.

[23] "Asymptotic stability and other properties of trajectories and transfer sequences leading to the bargaining set" (with G. Kalai and G. Owen), *International J. Game Theory* **4** (1975) 193–213.

[24] "Stable sets and stable points of set-valued dynamic systems with applications to game theory" (with B. Peleg), *SIAM J. Control and Optimization* **14** (1976) 985–995.

[25] "An Advantage of the bargaining set over the core," *J. Econ. Theory* **13** (1976) 184–192.

[26] "Playing an n-person game, an experiment in: *Contributions to Experimental Economics 8, Coalition Forming Behavior*, H. Sauermann, ed. J. C. B. Mohr (Paul Siebeck), Tübingen, 1978, pp. 231–328.

[27] "Geometric properties of the kernel, nucleolus and related solution concepts" (with B. Peleg and L. S. Shapley), *Math. of O.R.* **4** (1979) 303–338.

[28] "The super-additive solution for the Nash bargaining game" (with M. A. Perles), *International J. Game Theory* **10** (1981) 163–193.

[29] "Game theoretic analysis of a bankruptcy problem from the Talmud" (with R. J. Aumann), *J. Econ. Theory* **36** (1985) 195–213.

[30] "Bankruptcy games" (with I. J. Curiel and S. H. Tijs), *Zeitschrift für Operations Research* **31** (1987) 143–159.

[31] "Paths leading to the Nash set" (with G. Owen and B. Peleg), in: *The Shapley Value: Essays in Honor of Lloyd Shapley*, A. E. Roth, ed., Cambridge University Press, 1988, pp. 321–330.

[32] "The consistent Shapley value for hyperplane games," (with G. Owen), *International J. Game Theory* **18** (1989) 389–407.

[33] "Consistency in game theory and applications," in: *Game Theory and Applications*, Academic Press, 1990, pp. 183–186.

[34] "The consistent Shapley value for games without side payments" (with G. Owen), in: *Rational Interaction: Essays in Honor of John C. Harsanyi*, R. Selten, ed., Springer-Verlag, 1992, pp. 5–12.

[35] "The general nucleolus as a subset of the least core" (with J. A. M. Potters and S. H. Tijs), in: Proceedings of the International Conference on Game Theory, June 1991, Florence, MIT Press.

[36] "The general nucleolus and the reduced game property" (with J. A. M. Potters and S. H. Tijs), *International J. Game Theory* (1992) 85–106.

[37] "Monotonic games are spanning network games" (with A. van den Nouweland and S. H. Tijs), *International J. Game Theory* (1993) 419–427.

[38] "Credible equilibria in games with utilities changing during the play" (with J.-L. Ferreira and I. Gilboa), Submitted to *Games & Econ. Behavior*, 1992.

CONTENTS

A PREFACE TO THE AUMANN-SHAPLEY AND RUBINSTEIN PAPERS

In the Academic year 1975–6, all three of us were working on the model of repeated games. Rubinstein was a master's student at the Department of Mathematics at the Hebrew University, under the supervision of Bezalel Peleg. Shapley was at the Rand Corporation, where Aumann was consulting while spending a Sabbatical from the Hebrew University at Stanford. Simultaneously and independently, we proved two versions of the Perfect Folk Theorem for repeated games with complete information and the limit of the means. Rubinstein summarized his findings in Research Memorandum 26, Center for Research in Mathematical Economics and Game Theory, The Hebrew University of Jerusalem. Aumann and Shapley summarized their findings in a manuscript written in 1976; in 1978 it was distributed in connection with a one-day workshop on repeated games held at the Economics section of Stanford's Institute for Mathematical Studies in the Social Sciences. We had planned to write a joint paper but, being busy with other projects, failed to do so.

Rubinstein's memorandum included also a discussion of the perfect equilibria in a repeated game with the overtaking criterion, and of the strong perfect equilibrium in repeated games. Those sections were published in "Equilibrium in Supergames with the Overtaking Criterion", *J. Econ. Theory* 21 (1979) 1–9; and in "Strong Perfect Equilibrium in Supergames", *International J. Game Theory* 9 (1980) 1–12.

The outlined proof of the standard Folk Theorem in Section 3 of the Aumann-Shapley paper, and the example in Section 5, previously appeared in Appendices 2 and 3 of "Survey of Repeated Games", *Essays in Game Theory and Mathematical Economics in Honor of Oskar Morgenstern*, Wissenschaftsverlag, Bibliographisches Institut, Mannheim, Wien, Zurich, 1981, pp. 11–42.

We are very pleased at this opportunity to make this material, which appears here essentially in its original form, available to a wider public. It seems particularly appropriate for a volume dedicated to Michael Maschler, who has done so much to advance the theory of repeated games.

R. J. Aumann
A. Rubinstein
L. S. Shapley

Essay 1

Long-Term Competition—A Game-Theoretic Analysis

Robert J. Aumann[1] and Lloyd S. Shapley[2]

1. Introduction

There have been continuing expressions of interest from a variety of quarters in the development of techniques for modelling national behavior in a long-term context of continuing international rivalry — for short, "long term competition". The most characteristic feature of these models is that they extend over time in a fairly regular or repetitive manner. The underlying structure of possible actions and consequences remains the same, though parameters may vary and balances shift, and the decisions and policies of the national decision makers are by no means constrained to be constant or smoothly-varying, or even "rational" in any precisely identifiable sense. The use of game theory or an extension thereof is obviously indicated, and considerable theoretical progress has been made in this area. But the ability of the theory to handle real applications is still far from satisfactory. The trouble lies less with the *descriptive modelling,* — *i.e.*, formulating the "rules of the game" in a dynamic setting, than with the choice of a *solution concept* that will do dynamic justice to the interplay of motivations of the actors. (Game theoreticians, like mathematical economists, have always been more comfortable with static than dynamic models.) Since any predictions, recommendations, etc. that a mathematical analysis can produce will likely be very sensitive to the rationale of the solution that is used, and since the big difficulties are conceptual rather than technical, it seems both possible and worthwhile to discuss salient features of the theory without recourse to heavy mathematical apparatus or overly formal arguments, and thereby perhaps make the issues involved accessible to at least some of the potential customers for the practical analyses that we wish we could carry out in a more satisfactory and convincing manner.

[1]Department of Mathematics, The Hebrew University of Jerusalem, 91904 Jerusalem, Israel

[2]Department of Mathematics, UCLA, Los Angeles, CA 90024, U.S.A.

Two general types of "solution concept" are distinguished in game theory: *cooperative* notions, such as the core, bargaining set, von Neumann-Morgenstern stable sets, and Shapley value; and *noncooperative* notions, principally the Nash equilibrium point and its variants and elaborations, but including also the max-min solution based on "safety level" or "worst case" considerations. Cooperative notions are appropriate for situations where contracts among players are customarily adhered to and can be made legally binding; noncooperative notions where there is mistrust and no external enforcement mechanisms are available. The long term international scene is most naturally classified as noncooperative, since there is no effective international jurisdiction in most cases, even in the short run. Adherence to major international agreements is essentially a matter of national self-interest, and to be effective in the long run such agreements must be written to be self-enforcing, *i.e.*, so that it is to the continuing advantage of all sides to adhere to them.

Quite a bit is known about Nash noncooperative equilibria in "continuingly competitive" situations, and we shall review some of this material here. It turns out that individual self-interest in such situations can in fact dictate a kind of cooperative behavior, in many cases, sustained by the fear of "punishment" by the other players for failing to "cooperate" with the general plan — this in spite of the fact that the players have no way of legally binding themselves to carry out such punishment. The ability of the noncooperative theory to describe such arrangements and to account for their stability in a "selfish" world is an encouraging point in its favor. The price that is paid, however, is the high degree of nonuniqueness in the Nash solutions (as revealed in the two theorems described below), which removes from this theory most of its predictive power.

2. Repeated Game Models

In this section we shall review some of the known theory of a special kind of "continuingly competitive" game: that of *repeated games*. Given a finite game G in strategic form[3], we consider an infinite game G^*, each play of which consists of an infinite repetition of plays of G. In each play of G, or "round of G^*" the players are assumed to know the outcomes of all previous rounds. The payoff for G^* may be assumed to be of the limiting average

[3]a.k.a. "normal form."

form:[4]

$$\lim_{m\to\infty} \frac{1}{m} \sum_{t=1}^{m} h_t. \tag{2.1}$$

Here h_j is the payoff for G in the j'th round of G^*. Many authors call G^* the *supergame* of G.

An alternative form of the payoff for G^* involves discounting of future payoffs at a positive discount rate:

$$\sum_{t=1}^{\infty} (1-\rho)^t h_t. \tag{2.2}$$

For the time being we shall confine our discussion to the limiting average form (1), which treats the future as no less important than the present. Indeed, cumulatively, the future is all important in (1), since the contribution from any period of finite length will wash out in the long run. Nothing you actually do makes any difference; only your policies for the indefinite future have any significance. Despite this peculiarity, however, it should be remarked that for many purposes, both technical and conceptual, the limiting-average case behaves like the limit of the discounted-sum case, as the discount rate ρ in (2) goes to zero. Thus, used with care, the limiting average form can serve as an approximation to situations where a very low discount rate is appropriate. Long-term competition, almost by definition, would appear to fall into this category.

The basic theorem[5] about supergames states that a *necessary and sufficient condition for $h = (h^1, \ldots, h^n)$ to be the payoff vector of some Nash equilibrium point of G^* is that it be feasible and individually rational in G.*

Let us explain the key terms in this theorem. A "payoff vector" is simply an n-tuple of real numbers, where n is the number of players. The term "payoff vector" is used because the n coordinates signify the payoffs to the n players. By "feasible" we here mean "feasible in correlated strategies"; that is, a payoff vector is feasible if and only if it is in the convex hull of the set of payoff vectors that can be obtained by having the players play pure strategies. A payoff vector is called "individually rational" if each player receives at least his *min-max payoff*, which is the level of payoff below which he cannot be forced by the remaining players[6]. Finally, a "Nash equilibrium point," or "EP" is an n-tuple of strategies — one for each player in the

[4]A technical difficulty is that this limit need not always exist; this technical difficulty has a technical solution, which we do not wish to get involved with at present.

[5]This is a "folk-theorem"; it has never been published, but is well known to most workers in the field.

[6]When there are just two players, min-max=max-min.

game — such that each player's strategy is a best response to the $(n-1)$-tuple of the other players' strategies. In other words, no player can improve his own payoff by "defecting" to another strategy while the other players are held fixed.

To clarify the meaning of this theorem, let us see what it says about the well known "Prisoner's Dilemma." This is the two-player game whose strategic form is of the type given in the following table:

	Player II	
Player I	4, 4	0, 5
	5, 0	1, 1

The set of all feasible payoff vectors is indicated by the horizontally hatched region in Fig. 1, which is the convex hull of the four payoffs in the table. Since the minmax payoff to each player is 1, the set of all individually rational payoff vectors is indicated by the vertically hatched region. By the theorem, then, the set of payoff vectors arising from equilibrium points in the supergame is given by the *cross-hatched* region. Note in particular that the point (4,4) — the traditional "cooperative" outcome — appears as the payoff to an equilibrium point in the supergame of the Prisoner's Dilemma. The proof of the theorem is not difficult, and as the idea of the proof is important to a proper understanding of the situation we shall take a little space to outline it here. The "necessity" part is easily established; it is intuitively clear that equilibrium is not possible if any player is below his guaranteed minimum. The more interesting and significant part of the proof is the "sufficiency."

Assume for simplicity that $n = 2$ (there are just two players). Suppose h is a feasible, individually rational payoff vector. Here we may write

$$h = \sum_{m=1}^{k} \alpha_m h_m,$$

where the α are nonnegative weights that sum to 1 and the h_m are payoff vectors corresponding to pure strategy pairs in G. Suppose first that the α_m are rational numbers and express them in the form $\alpha_m = p_m/q$, where p_m are positive integers and q is their sum. The payoff vector h can then be achieved as a limiting average in G^* by having the players play for p_1 consecutive periods an n-tuple that achieves h_1, then for p_2 consecutive periods an n-tuple that achieves h_2, and so on; after q periods, we start again from the beginning.

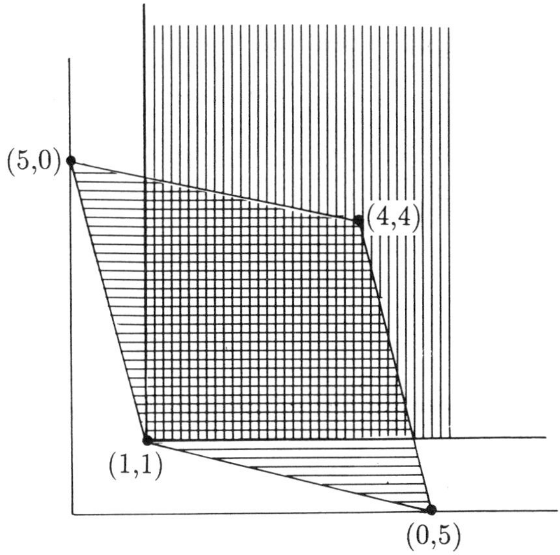

Figure 1: Feasible and individually rational payoffs in the Prisoner's Dilemma

If the α_m are irrational, the same effect can be attained by approximating to them by rational numbers, with increasing values of q, and playing once through each approximation in turn, to yield the desired limiting average.

This procedure, however, does not yet describe a Nash equilibrium point in G^*, and in fact does not even describe a pair of supergame strategies. It only describes a particular, feasible course of play. A supergame *strategy* must describe each player's responses to all possible actions of the other player, not only when he "plays along" with a prescribed course of play, such as the one described above, but also when he "defects". This is where the requirement that h be individually rational comes in.

Since h is individually rational, we have for each player i

$$h^i \geqq \max_{\sigma} \min_{\tau} H^i(\sigma, \tau),$$

where H^i is the payoff function to player i in the game G, σ ranges over all mixed strategies of player i, and τ ranges over all mixed strategies of the other player, j. By von Neumann's minimax theorem, there is a mixed strategy $\bar{\tau}$ of j such that for all mixed strategies σ' of i,

$$\min \max H^i(\sigma, \tau) \geqq H^i(\sigma', \bar{\tau});$$

hence in particular

$$h^i \geqq H^i(\sigma', \bar{\tau})$$

for all mixed strategies σ' of i. That means that by playing $\bar{\tau}$, j can hold i down to his max-min value, and *a fortiori* to h^i.

We may now describe an EP in G^* as follows: The players start by playing to obtain an average payoff of h, as outlined above. If at any stage a player i "defects" — *i.e.*, does not play the prescribed choice in G for that round — then starting from the next round, the other player j plays the mixed strategy $\bar{\tau}$ forever after. This will hold i's limiting average payoff down to at most h^i, so that he will have gained nothing by his defection. Thus, h is indeed the payoff to an EP.

3. Perfect Equilibrium Points

The above line of proof has been subjected to the following criticism: Though there is no question that the strategy pair as described constitutes an equilibrium point, it is not clear under what circumstances it would ever be used. In particular, it is possible that the strategy $\bar{\tau}$, while holding player i down to his minmax payoff, may also be very[7] disadvantageous to the player using it (or to one of the set of players participating in it, if $n > 2$). The equilibrium point dictates that $\bar{\tau}$ will continue to be played "forever," even if i defects only once. As we pointed out at the end of the previous section $\bar{\tau}$ is supposed to play the role of a deterrent. But an infinite unremitting repetition of $\bar{\tau}$ seems like an unreasonable response to a single act of defection, except for the fact that — in view of (1) — any finite period of "punishment" is no punishment at all. But by the same token, a single defection is also insignificant in the limit. Thus, the threatened response may still seem unreasonable, especially when, as is often the case, it is disadvantageous or costly to the user, and hence such an unremitting repetition may not be believable as a deterrent. To have a word for these EPs, let us call them "grim."

Let us try to pinpoint the dissatisfaction with grim EPs in a slightly more general framework. The "knowledge" that j (or, more generally $N \setminus \{i\}$) will respond to a defection on the part of i by an unrelenting stream of $\bar{\tau}$ is what keeps i from defecting; but if i does in fact defect, it may no longer be profitable for j (or $N \setminus \{i\}$) to respond with $\bar{\tau}$. This is what makes $\bar{\tau}$ unbelievable.

This kind of reasoning motivated a specialization of the notion of equilibrium point, first considered by R. Selten [2] and called by him a perfect equilibrium point[8]. To define this notion, we must recall more precisely the

---

[7]Compare [1].
[8]See also [3].

definition of a "strategy" for the player i in the supergame G^*. This is a function that tells i which pure G-strategy to choose on each round, as a function of what all the players, including i himself, did on all previous rounds. For each positive integer k, define G_k^* to be the "subgame" starting from the k'th period, *i.e.*, after $k-1$ rounds have been played, and continuing indefinitely from that point. Thus, $G^* = G_1^*$. Each n-tuple of strategies in G^*, together with a series of actual actions on the part of all players in the first $k-1$ rounds, induces an n-tuple of strategies in G_k^*. An n-tuple $\sigma = (\sigma_1, \ldots, \sigma_n)$ of strategies in G^* is called a *perfect equilibrium point* (or PEP) if for each k and for each series of actions of the players in the first $k-1$ periods of G^*, the induced n-tuple is an EP of G_k^*.

If we set $k = 1$ we see that a PEP is in particular an EP.

It's easy to see that a grim EP is in general *not* perfect, since if player i defects on round $k-1$ it will in general not be a best response in G_k^* for the other players to "punish" him; it may even be individually irrational. It thus appears that the notion of perfect equilibrium point might hold an answer to the problem of the believability of deterrents. In the next three sections we shall explore this matter somewhat further. To some extent our hopes turn out to be in vain: we shall find in the next section that the payoffs associated with PEPs in G^* are the same as those for grim EPs though the grim EPs themselves are excluded. The methods may be different, so to speak, and possibly more "believable" but the upshot is the same: there is no narrowing of the class of outcomes that can be sustained in equilibrium.

However, when we modify the payoff in G^* by introducing a positive discount rate (2), which we do in Sec. 5, we find that requiring "perfection" can significantly reduce the set of equilibrium outcomes. Moreover, the concept of believability does appear to play a significant role in the description of the perfect equilibria. Thus, the notion of perfection of equilibria, though not a panacea, does appear to give us a somewhat better handle on some of the problems that we wish to model.

4. Characterization of PEPs

This section is devoted to the following theorem:

Theorem 4.1. *The set of payoffs to perfect EPs in G^* coincides with the set of payoffs to ordinary EPs* — i.e., *it is the set of all feasible, individually rational payoffs in G.*

Again, to gain a good understanding of this theorem it is essential to outline the proof. As before, it is "sufficiency" that is the interesting part of the proof; the "necessity" follows from the previous theorem. We shall find the argument considerably more intricate than before.

To simplify the presentation we again assume that there are only two players. Moreover, in order to make the use of mixed strategies unnecessary, we shall assume that G is not in strategic form, with simultaneous choices by the two players, but is a game of perfect information with a single move for each player and no chance moves. Player I moves first, II is informed of I's move, and then II moves[9]. None of these assumptions are really required for the truth of the theorem, but they do simplify the proof.

Suppose h is a feasible, individually rational payoff vector of G. We shall describe a PEP with payoff h. As in Section 2, the description will be couched in terms of a tentative "agreement" on prescribed course of play. The agreement starts out as before with a sequence of choices which, when adhered to by both players, will lead to the desired limiting average payoff h. Let us call this "the cooperative sequence." Next, we shall specify how the players react to a defection — *i.e.*, a departure from the cooperative sequence by one of the players. In the previous proof, the reaction was unrelenting punishment. Here, instead, the PEP strategies will specify that a defection on the part of either player be punished by a sequence of choices by the other that forces the defector's average payoff down to within ε of his max-min value, where ε is a small number that may depend on the "date" of the defection. After the defector has thus been "beaten to within an inch of his life," the punisher relents and prescribed play returns to the cooperative sequence at the point of defection.

It should be noted, however, that not only are defections from the cooperative sequence punished, but also defections from any punishing sequence (in the subgame resulting from an earlier defection) are punished. A player who "should" punish and does not do so will himself be punished. This is what provides the motivation for the punisher actually to carry out the punishment, and so keeps the EP perfect.

The situation is a little complex; in order to convince ourselves that we have actually described a PEP we shall now give a more formal treatment. Without loss of generality we may assume that the number of choices available to I on each move is the same as the number of choices available to II; call the number m. Thus, $M = \{1, \ldots, m\}$ is the set of possible choices of the players at each move. When it's Player I's turn in the n-th round of G^* he has before him the full history of previous moves; this takes the form of a sequence $(x_1, \ldots, x_{2n-2})$, where $x_i \in M$ represents the choice made on the i'th move in G^* (it is I's or II's choice according as i is odd or even). Similarly when II must move, he has before him a sequence $(x_1, \ldots, x_{2n-1})$,

[9]Note that the apparent asymmetry of the players disappears in the supergame: in G^* the players move alternately, each with perfect information.

where again each $x_i \in M$. Let us call any finite sequence of members of M a *history*. A *strategy* for Player I [Player II] may be defined as a function from histories of even [odd] length to M. Thus, a pair of strategies is simply a function f from the set of *all* histories to M.

Now let $h = (h^I, h^{II})$ be the given feasible individually rational payoff vector. Let $(c_1, c_2, c_3, \ldots)$ be a fixed *cooperative sequence*, *i.e.*, a sequence of moves leading to the payoff h in G^*. Let $\bar{p}$ be a G-strategy for I (*i.e.*, a member of M) that holds II to his max-min payoff in G, and let $\bar{q}(\cdot)$ be a G-strategy for II (*i.e.*, a function from M to M) that holds I to his max-min payoff in G.

We wish to define a strategy-pair f for G^* which is a PEP and whose associated payoff is h. The definition of f will be inductive, based on the length k of the history on which it is being defined. On a history of length 0 we define f to be c_1; this simply means that the PEP will prescribe the choice of c_1 for the first move of Player I. Suppose now that f has been defined on all histories up to length $k-1$; we wish to define it on all histories of length k. Let $(x_1, \ldots, x_k)$ be such a history. We shall say that the ℓ'th move of that history $(1 \leqq \ell \leqq k)$ is a *defection* if $x_\ell \neq f(x_1, \ldots, x_{\ell-1})$. If $(x_1, \ldots, x_k)$ contains no defection, we define $f(x_1, \ldots, x_k) = c_{k+1}$. Otherwise, suppose the *most recent* defection in $(x_1 \ldots, x_k)$ occurred at move ℓ. If ℓ and $k+1$ have the same parity — *i.e.*, the player who is about to move is the same as the one who most recently defected — then we define $f(x_1, \ldots, x_k) = c_{k+1}$. If ℓ and $k+1$ have opposite parity, consider first the case in which ℓ is even, *i.e.*, Player II was the last to defect. In that case k is also even, so exactly $\frac{k}{2}$ rounds of G^* have now been completed and it is Player I's turn to move. Consider the average payoff of Player II as measured at the end of each of the rounds $\frac{\ell}{2}+1, \ldots \frac{k}{2}$, and let $\varepsilon_\ell = 1/\ell$. If any one of these averages is $\leqq \varepsilon_\ell +$ II's max-min value in G, then we define $f(x_1, \ldots, x_k) = c_{k+1}$; otherwise, we define $f(x_1, \ldots, x_k) = \bar{p}$. That means that Player I brings the offending Player II to within ε_ℓ of his max-min payoff and then returns to cooperative play.[10]

Finally, consider the case in which ℓ and $k+1$ have opposite parity and ℓ is odd, *i.e.*, Player I was the last to defect. In that case k is odd, $\frac{k-1}{2}$ rounds of G have been completed, Player I has already made his move in the $\frac{k+1}{2}$'th round, and it is now II's turn to move. Proceeding as before, we consider I's average payoff as measured at the end of each of the periods $\frac{\ell+1}{2}, \ldots, \frac{k-1}{2}$. If any of these averages was $\leqq \varepsilon_\ell +$ the max-min value to I in G, then we define $f(x_1, \ldots, x_k) = c_{k+1}$; otherwise, we define $f(x_1, \ldots, x_k) = \bar{q}(x_k)$. As before, that means that II brings the offending Player I to within ε_ℓ of his max-min value, then returns to cooperative play. (The difference is only

[10]Note that he never has occasion to look back beyond the most recent defection; the ε_j level of punishment suffices for all past transgressions.

that, because of the asymmetry in G, II's punishing move must depend on I's move in the same "round".)

This completes the formal description of the PEP that we described informally before; the reader should be able to convince himself that it is in equilibrium, is perfect, and yields the cooperative sequence $(c_1, c_2, \ldots, c_k, \ldots)$ with limiting average payoff h.

5. Discounted Payoffs in Repeated Games: Discussion of an Example

Thus far, we have been considering only the limiting average form of payoff for repeated games, corresponding intuitively to a future discount rate of zero. We shall now try to give an idea of how positive discount rates can affect the behavior of EPs and PEPs by studying an apparently simple but surprisingly revealing example.

Consider the following payoff matrix for G in strategic form, the players moving simultaneously:

		Player II	
		L	R
Player I	T	$0, 1$	$-p, -c+1$
	B	$1, 0$	$-p+1, -c$

Here, p and c are positive numbers ("punishment" and "cost"); we may think of them as being rather large. Thus, II may be in a position to damage I severely, but only at a cost to himself that may perhaps be unacceptable.[11]

In the repeated G^*, we shall use the discounted payoffs

$$\sum_{t=1}^{\infty} \alpha^t h_t^I \quad \text{and} \quad \sum_{t=1}^{\infty} \beta^t h_t^{II}$$

to Players I and II, respectively, where $0 < \alpha < 1$ and $0 < \beta < 1$. Sometimes we shall further assume that $\alpha \le \beta$, *i.e.*, that Player I has, if anything, a bigger discount rate (= shorter "utility horizon") than Player II.

[11] One could think of I and II as North Vietnam and the United States in the 1960's.

As is easily seen, the one-shot game G has a unique EP, namely (B, L), which yields the payoff $(1, 0)$. This means that the strategy-pair in which I always plays B and II always plays L (regardless of history) is a perfect EP of G^*, since obviously no defection, even in a subgame, can ever be profitable.

Player II, however, would naturally prefer the outcome $(0, 1)$, corresponding to (T, R). We shall now investigate under what conditions this outcome can be sustained by an EP, or by a PEP, in the discounted repeated game. Indeed, we shall find that it can be sustained by an EP if and only if $p \geqq 1/\alpha$; and, when $\alpha \geqq \beta$, that it can be sustained by a PEP if and only if $p \geqq 1/\alpha$ *and* $p/c \geqq (1-\beta)/\alpha\beta$. Thus, whereas the existence of an EP is independent of the cost of the punishment to the punisher (the parameter c), the existence of a PEP is not.

Let us first consider the EP question. We claim that the following "grim" strategy-pair:

$$\begin{cases} \text{I plays } T \text{ always} \\ \text{II plays } L \text{ so long as I plays } T, \text{ but plays } R \text{ forever} \\ \quad \text{if I ever plays } B \end{cases}$$

is an EP of G^*, provided that $p \geqq 1/\alpha$. Moreover, we claim that if $p < 1/\alpha$ there is no EP sustaining $(1, 1)$. Note first that II will certainly not wish to defect, as he cannot possibly improve on the sequence of payoffs $(1, 1, 1, \ldots)$. On the other hand, if I wishes to improve on his sequence $(0, 0, 0, \ldots)$, his best chance is to deviate to B at some time t_0, then keeping playing B forever. This yields him the payoff sequence $(0, \ldots, 0, 1, -p+1, \ldots)$, and this is worth

$$\alpha^{t_0}\left(\frac{1}{1-\alpha} - \frac{p\alpha}{1-\alpha}\right).$$

Since this is profitable to him *if and only if* $1 > p\alpha$ the truth of our claims is now evident.

For a numerical example, let $p = 2$. Then if $\alpha < .5$ there will be no EP sustaining the $(0, 1)$ outcome, as the rewards for defecting will outweigh any possible punishment. But if $\alpha \leq .5$, the strategy pair given above is clearly an EP.

Nothing in this result depends on the values of c or β. Yet, intuitively, one feels that the credibility of II's "threat," with which he extracts such a favorable outcome, ought to be *very* dependent on its cost. Our next object will be to show that a PEP that sustains the $(0, 1)$ outcome is not possible for large values of c.

First let us give an example of such a PEP. It happens that we can define it in a very simple way, making the instructions to the two players almost independent of the history:

1) In the first round, play (T, L).

2) If the choices in round $t-1$ were (T, L) or (B, R), play (T, L) in round t.

3) If the choices in round $t-1$ were (T, R) or (B, L), play (B, R) in round T.

The cooperative sequence resulting from this strategy pair is just a repetition of (T, L); this is worth $\beta/(1-\beta)$ to II and 0 to I. In checking for the PEP property, it is sufficient to look merely at deviations that occur in the first round of a typical subgame G_t^*. Suppose Player I defects when he is supposed to play T. His best possible payoff sequence form then on is $(1, -p+1, -p+1, \ldots)$, which is worth

$$\alpha^t \left(\frac{\alpha}{1-\alpha} - \frac{\alpha^2 p}{1-\alpha} \right)$$

to him. So if $p \geqq 1/\alpha$ he will not have any incentive to defect. Player II likewise will not defect when he is supposed to play T, as he cannot possibly improve on the payoff sequence $(1, 1, 1, \ldots)$. When I is supposed to play B, a defection could yield him at best the sequence $(-p, -p+1, 0, 0, \ldots)$; this is clearly inferior for all $p \geqq 0$ to the prescribed sequence $(-p+1, 0, 0 \ldots)$. Finally, when II is called upon to play 2 at the beginning of G_t^*, he will have to compare his prescribed payoff sequence $(-c, 1, 1, 1, \ldots)$ with sequences like $(0, -c, 1, 1, 1, \ldots)$, $(0, 0, -c, 1, 1, 1, \ldots)$, etc., which he can obtain by defecting for 1,2, etc. rounds, or even the sequence $(0, 0, 0, \ldots)$ which he can obtain by perpetually defecting. In the discounted sum, these "heresies" are worth

$$\beta^k(-c\beta + \beta^2/(1-\beta)),\ k = t+2,\ t+3,\ \ldots,$$

or 0, while "orthodoxy" is worth

$$\beta^t(-c\beta + \beta^2/(1-\beta)).$$

So if $c \leqq \beta(1-\beta)$, II cannot gain by defecting. The given strategy pair is therefore a PEP on the assumptions that $p \geqq 1/\alpha$ and $c \leqq \beta/(1-\beta)$, as diagrammed in Figure 2.

To wrap up our example, it is necessary to show that there are significant cases, where, because of the positive discount rate, a PEP does *not* exist. Showing nonexistence is a more difficult undertaking, because in general a PEP can be a very complex thing. In particular, while pure G-strategies have sufficed up to now, we cannot ignore the possible use of mixed strategies against a defection. In our example, if p and c are both large numbers, the threat of a small probability of using R may be enough to keep I in line while holding the (expected) cost to a level that II can accept[12]. However, it would be out of place in this discussion to develop the elaborate technical apparatus of mixed strategies just for the sake of one example, whose purpose is only illustrative. Instead, we shall adopt a far simpler expedient, called "convexification in pure strategies," which is more or less equivalent to the introduction of mixed strategies.

In our example, this convexification merely means allowing Player II the option of "scaling down" his punishment by giving him a continuum of strategies in G, as follows:

		Player II	
		L	$R\lambda$ $(0 < \lambda \leqq 1)$
Player I	T	$0, 1$	$-\lambda p, -\lambda c + 1$
	B	$1, 0$	$-\lambda p + 1, -\lambda c$

Here, $\lambda = 1$ corresponds to the old R and $\lambda = 0$ corresponds to the old L. (However, we still indicate the latter choice by a separate column in the matrix. Playing $R\lambda$ has much the same effect as playing a mixed strategy $\{R$ with probability λ, L with probability $1 - \lambda\}$, and it can be shown (though we shall not do it here) that if the new G^* has no PEP in *pure* strategies that sustains the cooperative sequence $((T, L), (T, L), \ldots)$, then the original G^* (with the same values of α, β, p, c) has no PEP in pure or mixed strategies that sustains that sequence.

Consider now a play of the revised G^*, with II making the sequence of moves $\Lambda = (\lambda_1, \lambda_2, \ldots, \lambda_t, \ldots)$. The total punishment received by I is then given by $P = P(\Lambda) = \sum_{t=1}^{\infty} \lambda_t \alpha^t p$, and the total cost incurred by II is $C = C(\Lambda) = \sum_{t=1}^{\infty} \lambda_t \beta^t c$. We now bring in the assumption, not used until now, that $\alpha \leqq \beta$. This implies that $\sum_{t=1}^{\infty} \lambda_t(\beta^t - \alpha^t) \geqq 0$, so that $P/C \leqq p/c$. This inequality shows that it is most efficient, in terms of the damage/cost ratio, for II to punish *immediately*; he thereby minimizes his cost for a given level of deterrence. It follows that the game has a PEP, of the type described above, whenever there is any number λ such that λp and λc satisfy the inequalities

$$\lambda p \geqq 1/\alpha, \qquad \lambda c \leqq \beta/(1 - \beta).$$

[12]This is a realistic consideration for the world of nuclear politics and arms races, where the pressure of the nuclear deterrent is felt in every situation that creates any perceptible risk that the situation might escalate out of control.

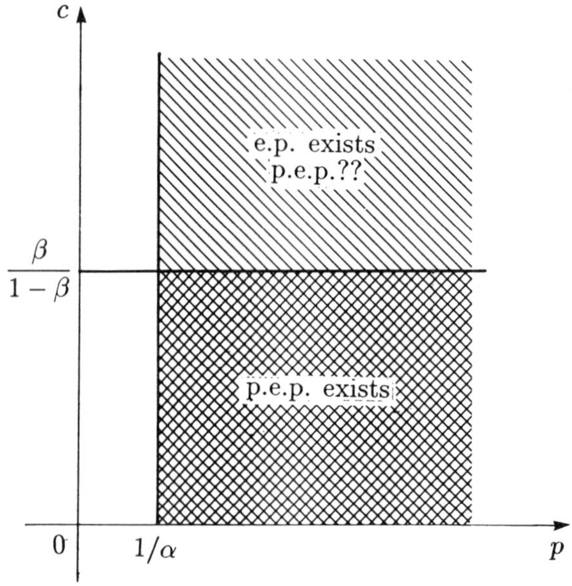

Figure 2

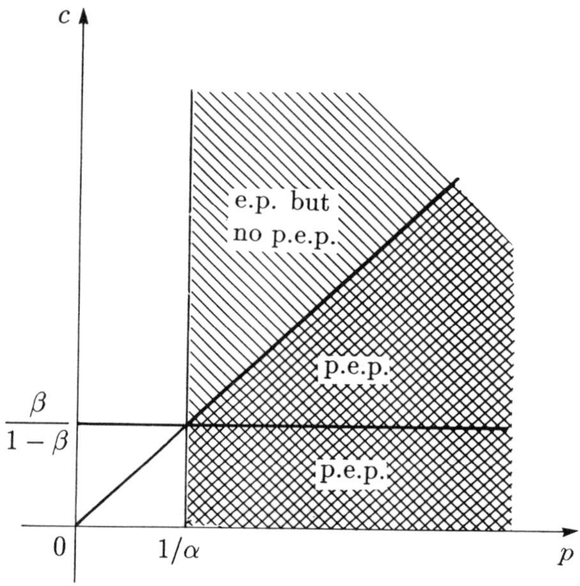

Figure 3

This is illustrated in Fig. 3. As we can see, there is a critical ratio of c to p, namely

$$R_0 = \frac{\alpha\beta}{1-\beta},$$

above which no such λ can be found.

For a numerical example, if $\alpha = \beta = .75$ then $R_0 = 2.25$. If $p = 100$ and $c = 200$ then $c/p < R_0$, and we may, for example, choose $\lambda = .014$, giving us $\lambda c = 2.8 < \beta/(1-\beta) = 3$ and $\lambda p = 1.4 > 1/\alpha = 1.33$. So a 1.4 percent chance of II using his threat strategy R after a defection by I sustains the perfect equilibrium at $(0, 1)$.

We can also make the converse argument. If (c, p) is *not* in the cross-hatched region indicated in Fig. 3, then there is no way for II to inflict any given amount P of punishment without incurring a cost of more than R_0P. By the foregoing, it is clear that this is above the cost that he can "afford"; in other words, he would prefer to accept his max-min payoff of 0 forever, rather than carry out the requisite threat. So a PEP cannot exist.

6. References

[1] R. J. Aumann, "Acceptable Points in General Cooperative n-Person Games," in *Contributions to the Theory of Games IV*, Annals of Math. Study **40**, Princeton University Press, 1959, pp. 287–324.

[2] R. Selten, "Spieltheoretische behandlung eines oligopolmodells mit nachfrageträgheit," *Zeitschrift für die gesamte Staatswissenschaft* **12** (1965) 301–324.

[3] R. Selten, "Re-examination of the perfectness concept for equilibrium points in extensive games," *International J. Game Theory* **4** (1975) 25–55.

Essay 2

Equilibrium in Supergames

Ariel Rubinstein[1]

This paper is the first part of Research Memorandum 25, The Center for Research in Game Theory and Mathematical Economy which was published in May 1977 and was based on my M.Sc. Thesis, written in 1975-6 at the Hebrew University of Jerusalem under the supervision of Professor B. Peleg. Other parts of the report were published in "Equilibrium in Supergames with the Overtaking Criterion," *J. Econ. Theory* **21** (1979) 1–9, and "Strong Perfect Equilibrium in Supergames," *International J. Game Theory* **9** (1980) 1–12. The main theorem in this paper was discovered simultaneously by R. J. Aumann and L. S. Shapley.

1. Introduction

There are significant differences between the situation of players undertaking to play a single game, and players who know that they will play the same game repeatedly in the future. Strategy in the first case is a single action; in the second, it is a sequence of rules, each one of which pertains to the outcomes preceding it. The preferences of the participants are determined partly by temporal considerations. The participants may adopt risky strategies, "supported" by threats of retribution in the future.

Analysis of a finite sequence of identical games shows that this model is inadequate for the analysis. If the number of games is finite and known initially, the players will treat the last game as if it were a single game. As the threats implicit in the game before last are proven to be false threats, the game before last is treated as a single game, and so on. (For a detailed analysis, see Luce and Raiffa [8].)

In order to avoid "end-points" in the model, we define a "supergame" as an infinite sequence of identical games, together with the players' evaluation relations (that is, their preference orders on utility sequences). Obviously, the assumption of an infinite planning horizon is unrealistic, but it is an approximation to the situation we wish to describe (see Aumann [1]).

The literature deals mainly with comparison of equilibrium concepts in supergames and single games. (See Aumann [1] and [2]. The results are

[1]Department of Economics, Tel-Aviv University, Tel Aviv, Israel

simply in [5]; see also [3] and [4].) Other papers emphasize e concept of supergames in economics are Friedman [6] and

s paper, as in most of the literature, it is assumed that the partic- s evaluate the utility flows according to the criterion of the limit of the means of the flows. The drawback in this evaluation relation is that it ignores any finite time interval.

The formal model, described in Section 2, is taken from Roth.[2]

The single game is given in strategic form. A Nash equilibrium in an n-player super game is an n-tuple of supergame strategies, such that no player may singly deviate profitably from his strategy. A steady Nash equilibrium is one which produces identical outcomes for every game played. In Section 3, the steady equilibrium points will be characterized by a "two-stage" finite game in which the time element is reduced to "present" and "future."

An equilibrium point will be called perfect if after any possible "history," the strategies planned are an equilibrium point. In other words, no player ever has a motive to change his strategy. The main theorem of this paper provides a complete characterization of steady perfect equilibrium outcomes for supergames with the limit of means evaluation relation. It is proved that the requirement of perfection does not alter the set of steady equilibrium outcomes.

2. The model

(i) The *single* game G is a game in strategic form

$$G = \langle \{S_i\}_{i=1}^n, \{\pi_i\}_{i=1}^n \rangle .$$

The set of players is $N = \{1, \dots, n\}$. For each $i \in N$, the set of strategies of i is S_i; S_i is assumed to be non-empty and compact. $S = \prod_{i=1}^n S_i$ is the set of outcomes. An element in S will be called an outcome of G. The preference relations of player i are defined by utility function $\pi_i : S \to \Re$ (where $\Re$ is the set of the reals), which are continuous in the product topology.

Given $\sigma \in S$, a payoff vector is the n-tuple $\pi(\sigma) = \langle \pi_1(\sigma), \dots, \pi_n(\sigma) \rangle$. For convenience we will denote the $(n-1)$-tuple $\langle \sigma_1, \dots, \sigma_{i-1}, \sigma_{i+1}, \dots, \sigma_n \rangle$ by σ_{-i}, and the n-tuple σ by $\langle \sigma_{-i}, \sigma_i \rangle$. σ will be called a *(Nash) equilibrium* if for all i and for all $s_i \in S_i$, $\pi_i(\sigma_{-i}, s_i) \le \pi_i(\sigma)$.

[2] I wish to thank Professor A. E. Roth for permitting me to use the model described in Roth [9].

If the set of strategies is finite, and it is possible to adopt mixed strategies, we can identify S_i with the set of mixed strategies of i, and π_i with the expected payoff.

(ii) The *supergame* G^∞ is the game $\langle G \prec_1, \ldots, \prec_n \rangle$, where G is a single game and the $\prec_i$ are evaluation relations on real number sequences; that is, $\prec_i$ is a binary relation on $\pi_i(S)^N$ ((A^N is the set of sequences of elements in A) where $\pi_i(S)$ is the range of π_i on S. $\prec_i$ will be transitive and anti-symmetric but not necessarily a total order.

The set of outcomes at time t, $S(t)$, is S. A strategy for i in G^∞ is a set $\{f_i(t)\}_{t=1}^{\infty}$, where $f_i(1) \in S(1)$, and for $t \geq 2$, $f_i(t) : \prod_{j=1}^{t-1} S(j) \to S_i$. Thus, a supergame strategy is a choice of strategies at every stage, possibly dependent on the outcomes preceding the choice. We assume all players know all the choices made in the past by all the players.

The set of strategies of i will be denoted by F_i. F is the set of n-tuples of the strategies; $F = \prod_{i=1}^{n} F_i$.

Given $f \in F$, the outcome at time t will be denoted by $\sigma(f)(t)$, and is defined inductively by

$$\sigma(f)(1) = (f_1(1), \ldots, f_n(1))$$
$$\sigma(f)(t) = (\ldots, f_i(t)(\sigma(f)(1), \ldots, \sigma(f)(t-1)), \ldots) .$$

We will define an evaluation relation $\bar{\prec}_i$ on F, induced by $\prec_i$, as follows.

$$(\forall f, g \in F)\,(f \bar{\prec}_i g \;\Leftrightarrow\; \{\pi_i(\sigma(f)(t))\}_{t=1}^{\infty} \prec_i \{\pi_i(\sigma(g)(t))\}_{t=1}^{\infty}) .$$

Given $f \in F$, we will denote $(f_1, \ldots, f_{i-1}, f_{i+1}, \ldots, f_n)$ by f_{-i}. $f \in F$ is a (Nash) equilibrium in the supergame G^∞ if there is no $h_i \in F_i$ such that $f \bar{\prec}_i (f_{-i}, h_i)$. $f \in F$ is steady if there exists $\sigma \in S$ such that for all t, $\sigma(f)(t) = \sigma$. If $f \in F$ is steady, we will denote the corresponding σ by $\hat{\sigma}(f)$.

3. Characterization of steady Nash equilibria

We will assume that each player is characterized by a single evaluation relation. We will merely assume that the evaluation relations are *reasonable*, in the sense that they satisfy:

(A.1) If for all t, $y_t \equiv y_0$, and $x_t \equiv x_0$, then $x_0 < y_0$ implies $x \prec y$.

(A.2) If $z \prec x$, and $x \leq y$ (that is, for all $t \in N$, $x_t \leq y_t$), then $z \prec y$.

(A.3) If there exists an $a \in A$ such that $(a, x_1, x_2, \ldots) \prec (a, y_1, y_2, \ldots)$, then $x \prec y$.

EXAMPLES:

1 The *limit of means* evaluation relation is defined by

$$x \prec y \text{ iff } \varliminf_{n\to\infty} \frac{1}{n}\sum_{t=1}^{n} x_t < \varliminf_{n\to\infty} \frac{1}{n}\sum_{t=1}^{n} y_t \,.$$

2 *Overtaking criterion* evaluation relation, defined by

$$x \prec y \text{ iff } \varliminf_{n\to\infty} \sum_{i=1}^{n}(y_i - x_i) > 0 \,.$$

3 The evaluation relation with *discount* parameter $0 < \delta < 1$, defined by

$$x \prec y \text{ iff } \sum_{t=1}^{\infty} \delta^t x_t < \sum_{t=1}^{\infty} \delta^t y_t \,.$$

This section has two goals. The first is to characterize the steady Nash equilibria using the equilibria of a "finite" two-stage game, G^2, derived from G^∞. The game G^2 is a twofold repetition of G. A strategy in G^2 contains decisions about the "present," the first game, and the "future," the second game. The latter decision depends on the outcome of the former. The second aim is to show that the considerations included in the notion of equilibrium of a supergame can be studied by a two-stage model.

We will now define the derived game, G^2. A strategy for a player i in G^2 is a pair, $\langle f_i(1), f_i(2)\rangle$, where $f_i(1) \in S_i$ and $f_i(2) : S(1) \to S_i$. We will denote the strategies of i by F_i^2, and write $F^2 = \prod_{i\in N} F_i^2$.

We define a partial order $\prec_i^2$ on $\pi_i(S) \times \pi_i(S)$ as follows.

$$(b_1, b_2) = b \prec_i^2 a = (a_1, a_2) \quad \text{iff} \quad \begin{cases} 1)\ b_1 < a_1 \text{ and} \\ 2)\ (b_1, b_2, b_2, b_2, \ldots) \prec_i (a_1, a_2, a_2, a_2, \ldots), \end{cases}$$

where $\prec_i$ is the evaluation relation of a player i in G^∞.

The outcome of G^2, where a player adopts strategy $f \in F^2$ is defined by:

$$\sigma(f)(1) = (f_1(1), \ldots, f_n(1))$$
$$\sigma(f)(2) = (f_1(\sigma(f)(1)), \ldots, f_n(\sigma(f)(1))) \,.$$

$f \in F^2$ will be called *steady* if there exists a $\sigma \in S$ such that $\sigma(f)(1) = \sigma(f)(2) = \sigma$. Such a σ will be denoted by $\hat{\sigma}(f)$. $f \in F^2$ will be called an *equilibrium* if there is no i and $g_i \in F_i^2$ such that

$$(\pi_i(\sigma(f)(t)))_{t=1}^2 \prec_i^2 (\pi_i(\sigma(f_{-i}, g_i)(t)))_{t=1}^2 \,.$$

EXAMPLES:

1) If $\prec$ is the limit of means evaluation relation, the relation induced is:

$$b \prec^2 a \text{ iff } b_1 < a_1 \text{ and } b_2 < a_2 \ (a, b \in \Re^2) .$$

2) If $\prec$ is the overtaking evaluation relation,

$$b \prec^2 a \text{ iff } b_1 < a_1 \text{ and } b_2 \leq a_2 .$$

3) If $\prec$ is the evaluation relation with discount parameter $0 < \delta < 1$, then

$$b \prec^2 a \text{ iff } b_1 < a_1 \text{ and } b_1 + \frac{\delta}{1-\delta} b_2 < a_1 + \frac{\delta}{1-\delta} a_2 .$$

Thus, a player deviating at any one time only if he will derive certain profit in the foreseeable future, behaves according to the evaluation relation induced in G^2 by the limit of means relation. A player deviating only if he will not "lose" in the future, behaves according to the relation in G^2 induced by the overtaking criterion. If there exists $\epsilon > 0$, such that a player will deviate in the future iff the difference between his present profit and future loss exceeds ϵ, (here $\frac{a_1 - b_1}{b_2 - a_2} > \epsilon$ iff $a_1 + \epsilon a_2 > b_1 + \epsilon b_2$); then, the evaluation relation in G^2 corresponding to this behavior is that induced by the evaluation relation with discount parameter $\delta = \frac{\epsilon}{1+\epsilon}$.

Remark 3.1. For a supergame $G^\infty = \langle G, \prec_1, \ldots, \prec_n \rangle$, where for all i, the evaluation relations are reasonable, and where an equilibrium σ exists in G, define $f \in F$ by: "For all i and for all t, $f_i(t) \equiv \sigma_i$." Clearly, f is a steady equilibrium in G^∞ (even perfect; see the definition in Section 4), satisfying $\hat{\sigma}(f) = \sigma$. Thus, in a supergame where the single game has an equilibrium, we are guaranteed the existence of an equilibrium.

Proposition 3.2. *Let $G^\infty = \langle G, \prec_1, \ldots, \prec_n \rangle$ be a supergame where the $\prec_i$ are reasonable evaluation relations. If there exists $g \in F^2$, a steady equilibrium in G^2, the derived game, such that $\hat{\sigma}(g) = \sigma$, then there exists $f \in F$, a steady equilibrium in G^∞ such that $\hat{\sigma}(f) = \sigma$.*

Proof: Given $s \in S$, let $r_i(s) \in S_i$ satisfy

$$\pi_i(r_i(s), s_{-i}) = \max_{t_i \in S_i} \pi_i(t_i, s_{-i}) .$$

Since g is an equilibrium in G^2, there exist $\gamma^i \in S$ satisfying

$$(\pi_i(\sigma), \pi_i(\sigma)) \not\prec_i^2 (\pi_i(r_i(g(1)), g_{-i}(1)), \pi_i(r_i(\gamma^i), \gamma_{i-i})) .$$

We may assume that if $\pi_i(r_i(\sigma), \sigma_{-i}) \leq \pi_i(\sigma)$, then $\gamma^i = \sigma$, and we use the notation $[a] = (a, a, a \ldots)$, and $(a, [b]) = (a, b, b, b, \ldots)$.

Define $f_i \in F_i$ for $i \in N$ as follows.

$$f_i(1) = \sigma_i$$

$$f_i(t)(s(1), \ldots, s(t-1)) = \begin{cases} \gamma_i^j & \text{if there exists } T \leq t-1 \text{ such that:} \\ & s(1) = \cdots = s(T+1) = \sigma, \\ & s_{-j}(T) = \sigma_{-j}, \text{ and } s_j(T) \neq \sigma_j \\ \sigma_i & \text{otherwise} \end{cases}$$

Then $\hat{\sigma}(f) = \sigma$ and f is a G^∞ equilibrium, since if $(f_{-i}, h_i) \bar{\succ}_i f$:
1) If $\pi_i(\sigma) \geq \pi_i(r_i(\sigma), \sigma_{-i})$, then $\pi_i(\sigma) \geq \pi_i(\sigma(g_{-i}, h_i)(t))$ for all t, and thus, according to (A.2) it will follow that $[\pi_i(\sigma)] \succ_i [\pi_i(\sigma]$ contradicting irreflexivity.
2) If $\pi_i(\sigma) < \pi_i(r_i(\sigma), \sigma_{-i})$, then let t_0 be the minimum satisfying $h_i(t_0)(\sigma, \sigma, \ldots, \sigma) \neq \sigma_i$. If $\{\pi_i(\sigma(f_{-i}, h_i)(t))\}_{t=1}^{\infty} \succ_i [\pi_i(\sigma)]$, then repeated applications of (A.3) yield

$$\{\pi_i(\sigma(f_{-i}, h_i)(t))\}_{t=t_0}^{\infty} \succ_i [\pi_i(\sigma)] .$$

But $\pi_i(r_i(\sigma), \sigma_{-i}) \geq \pi_i(\sigma(f_{-i}, h_i)(t_0))$ and for $t > t_0$, $\pi_i(r_i(\gamma^i), \gamma^i_{-i}) \geq \pi_i(\sigma(f_{-i}, h_i)(t))$. Applying (A.2),

$$(\pi_i(r_i(\sigma), \sigma_{-i}), [\pi_i(r_i(\gamma^i), \gamma^i_{-i})]) \succ_i [(\pi_i(\sigma)] ,$$

contradicting the choice of γ^i. ∎

Proposition 3.3. *Let $G^\infty = \langle G, \prec_1, \ldots, \prec_n \rangle$ where $\prec_i$ are reasonable evaluation relations. If $f \in F$ is a steady equilibrium in G^∞, and $\hat{\sigma}(f) = \sigma$, then there exists a G^2 stationary equilibrium $g \in F^2$, such that $\hat{\sigma}(g) = \sigma$.*

Proof: Suppose not. Then there is an i and $\tau_i \in S_i$ satisfying:

1) $\pi_i(\sigma_{-i}, \tau_i) > \pi_i(\sigma)$.

2) For all $s \in S$, there are $t_i \in S_i$ such that

$$\pi_i(\sigma_{-i}, \tau_i), [\pi_i(t_i, s_{-i})] \succ_i [\pi_i(\sigma)] .$$

Thus, applying (A.2),

$$\left(\pi_i(\sigma_{-i}, \tau_i), \left[\min_{s_{-i}} \max_{t_i} \pi_i(t_i, s_{-i})\right]\right) \succ_i [\pi_i(\sigma)] .$$

Define h_i, a strategy in G^∞, by

$$h_i(1) = \tau_i$$

$$h_i(t)(s(1)\ldots s(t-1)) = r_i(f(t)(s(1)\ldots s(t-1)))\ .$$

Applying (A.2), we obtain

$$\{\pi_i(\sigma(f_{-i}, h_i)(t))\}_{t=1}^{\infty} \succ_i [\pi_i(\sigma)]\ .$$

in contradiction to f being an equilibrium. ▮

Definition 3.4. $s \in S$ is a *weakly forced outcome*[3] in G if, for every i, there is an $r \in S$ such that for all $t_i \in S_i$, $\pi_i(r_{-i}, t_i) \leq \pi_i(s)$.

Thus, in a weakly forced outcome, each player's payoff is at least as large as the punishment the other players can inflict on him, that is, at least $\min_{r \in S} \max_{t_i \in S_i} \pi_i(r_{-i}, t_i)$.

Example 3.5. Let S_i be the set of mixed strategies of i, $i = 1, 2$. In a matrix game with a payoff matrix

2,2	0,3
3,0	1,1

π_i is the expected payoff of i.

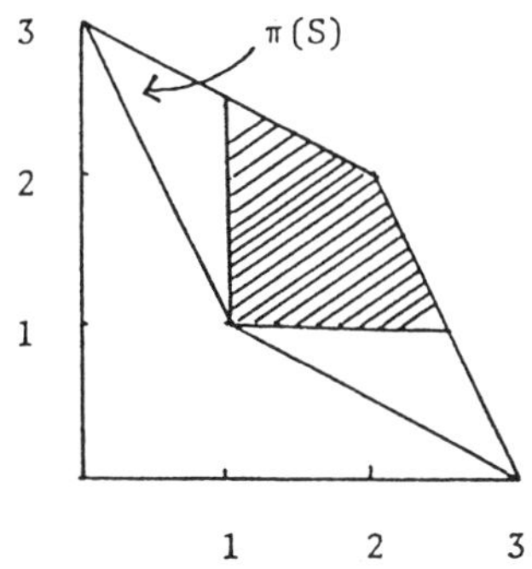

[3]The term "individually rational" is by now the established terminology for this concept.

Proposition 3.6. *If $\prec_i$ is the limit of means evaluation relation in G^∞, the steady outcomes of steady equilibria are the weakly forced outcomes.*

Proof: There exists $g \in F$, a steady G^∞ equilibrium, such that $\hat{\sigma}(g) = \sigma$, iff there exists $f \in F^2$, a steady G^2 equilibrium such that $\hat{\sigma}(f) = \sigma$ and this holds iff for no $h_i \in F_i^2$, $(\sigma, \sigma) \not\prec_i^2 (\sigma(h_i, f_{-i})(t))_{t=1}^2$, *i.e.*, iff for all i, either $\pi_i(r_i(\sigma), \sigma_{-i}) \leq \pi_i(\sigma)$ or

$$\pi_i(r_i(\gamma^i), \gamma^i_{-i}) = \underline{\lim}\, \tfrac{1}{n}\left[\pi_i(r_i(\sigma), \sigma_{-i}) + (n-1)\pi_i(r_i(\gamma^i), \gamma^i_{-i})\right] \leq \pi_i(\sigma)$$

and this holds iff for every i, there exists $\gamma^i \in S$ such that $\pi_i(r_i(\gamma^i), \gamma^i_{-i}) \leq \pi_i(\sigma)$. ∎

Proposition 3.7. *Let $G^\infty = \langle G, \prec_1, \ldots, \prec_n \rangle$ be a supergame with reasonable evaluation relations. A necessary condition for $f \in F$ to be a steady equilibrium is that $\hat{\sigma}(f)$ is a weakly forced outcome.*

Proof: By Proposition 3.3, there exists $g \in F$, a steady equilibrium in G^2 such that $\hat{\sigma}(g) = \hat{\sigma}(f)$. If $\hat{\sigma}(f)$ is not weakly forced, there exists i such that for all $s \in S$, $\pi_i(r_i(s), s) > \pi_i(\hat{\sigma}(f))$. S is compact. Thus, there exists $\epsilon > 0$ such that for all $s \in S$ $\pi_i(r_i(s), s_{-i}) > \pi_i(\hat{\sigma}(f)) + \epsilon$. Define $h_i \in F_i^2$ by:

$$h_i(1) = r_i(\hat{\sigma}(f))$$

$$h_i(2) = r_i(g(1)(\hat{\sigma}_{-i}(f), r_i(\hat{\sigma}(f)))).$$

Applying (A.1), we obtain:

$$(\pi_i(\sigma(g_{-i}, h_i)(t))_{t=1}^2 \succ_i^2 (\pi_i(\hat{\sigma}(f)), \pi_i(\hat{\sigma}(f)))) ,$$

in contradiction to g being a G^2 equilibrium. ∎

4. Perfect equilibria

The definition of equilibrium as in Section 2 was shown in Section 3 to be weak. The set of equilibria is too large, and it is natural to introduce further reasonable restrictions to obtain a stronger characterization. One possible requirement is that a deviation will prove unprofitable to a player at all stages of the game, not only at the beginning; thus, under no circumstances will be induced to change his original strategy.

Definition 4.1. $f \in F$ is a *perfect equilibrium* if for all $r(1), \ldots, r(T) \in S$, the strategy profile $\bar{f}$ defined by

$$\bar{f}_i(t)(s(1), \ldots, s(t-1)) = f_i(t+T)(r(1), \ldots, r(T), s(1), \ldots, s(t-1))$$

is an equilibrium.

Not only is it unprofitable for one player to alter his strategy, but no player can perform manipulative maneuvers since after each "history," all players prefer not to deviate.

The following proposition characterizes the steady perfect equilibrium in a supergame with evaluation relations determined by the limit of means criterion. A similar result was discovered independently by Aumann and Shapley.

Proposition 4.2. *If $\sigma \in S$, there is a perfect steady equilibrium $f \in F$ such that $\hat{\sigma}(f) = \sigma$, iff σ is a weakly forced outcome.*

Proof: Necessity follows from Proposition 3.6. Let $\sigma \in S$ be a weakly forced outcome. Let γ^i be strategies satisfying

$$\max_{t_i \in S_i} \pi_i \left(\gamma^i_{-i}, t_i\right) \le \pi_i(\sigma)$$

((γ^i will be the strategy for punishing player i). Define $f_i(t+1)(s(1), \ldots, s(t))$ and $P(s(1), \ldots, s(t))$ inductively as follows. ($P(s(1), \ldots, s(t))$ is interpreted as the set of players deserving punishment after the history $(s(1), \ldots, s(t))$:

$$P(\emptyset) = \emptyset, \quad \text{and} \quad f_i(1) = \sigma_i \ ,$$

$$P(s(1), \ldots, s(t)) = \begin{cases} \{i\} & \text{if } s_{-i}(t) = \gamma^i_{-i},\ P(s(1), \ldots, s(t-1)) = \{i\}, \\ & \quad \text{and } \frac{1}{t}\sum_{k=1}^{t} \pi(s(k)) \ge \pi(\sigma) + \frac{1}{\sqrt{t}} \\ \{i\} & \text{if } P(s(1), \ldots, s(t-1)) = \emptyset \ \&\ s_{-i}(t) = \sigma{-i} \\ & \text{but } s_i(t) \ne \sigma_i \ \&\ \frac{1}{t}\sum_{k=1}^{t} \pi_i(s(k)) \ge \pi_i(\sigma) + \frac{1}{\sqrt{t}} \\ \emptyset & \quad \text{otherwise} \end{cases}$$

$$f_i(t+1)(s(1), \ldots, s(t)) = \begin{cases} \gamma^j_i & \text{if } j \ne i \text{ and } P(s(1), \ldots, s(t)) = \{j\} \\ \sigma_i & \quad \text{otherwise} \end{cases}$$

Let $r(1), \ldots, r(T) \in S$; denote

$$\hat{f}_i(1) = f_i(T+1)(r(1), \ldots, r(T))$$

$$\hat{f}_i(t)(s(1), \ldots, s(t-1)) = f_i(t+T)(r(1), \ldots, r(T), s(1), \ldots, s(t-1)) \ .$$

We will show that $\hat{f}$ is an equilibrium. The following two lemmas will complete the proof. ∎

Lemma 4.3. *There exists T_1 such that for all $t \geq T_1$, $\sigma(\hat{f})(t) = \sigma$.*

Proof: If $P(r(1),\ldots,r(T)) = \emptyset$, then $\sigma(\bar{f})(t) = \sigma$ for all $t \geq 1$. If $P(r(1),\ldots,r(T)) = \{j\}$, then

$$\begin{aligned}\frac{1}{T+t}\left(\sum_{k=1}^{T}\pi_j(r(k)) + t\pi_j(\gamma^j)\right) &\leq \frac{1}{T+t}\left(\sum_{k=1}^{T}\pi_j(r(k)) + t\pi_j(\sigma)\right)\\ &\leq \frac{1}{T+t}\sum_{k=1}^{T}\pi_j(r(k)) + \pi_j(\sigma)\\ &< \frac{1}{\sqrt{T+t}} + \pi_j(\sigma)\end{aligned}$$

for sufficiently large t. ∎

Lemma 4.4. *Let $h_i \in F_i$. For every t_0, there is $t \geq t_0$ such that*

$$\frac{1}{t}\sum_{k=1}^{t}\pi_i(\sigma(\hat{f}_{-i},h_i)(k)) < \pi_i(\sigma) + \frac{1}{\sqrt{t}}\,.$$

Proof: It suffices to consider the case $t_0 \geq T_1$, where T_1 is given by Lemma 4.3; note also that $t \geq T_1$ implies $P(s(1),\ldots,s(t)) \subseteq \{i\}$. If for all $t \geq T_0$, i does not deserve punishment after $\{\sigma(\hat{f}_{-i},h_i)(k)\}_{k=1}^{t}$, then $\sigma(\hat{f}_{-i},h_i) = \sigma$ for all $t_0 < k$, and

$$\frac{1}{t}\sum_{k=1}^{t}\pi_i(\sigma(\hat{f}_{-i},h_0)(k)) = \frac{1}{t}\sum_{k=1}^{t_0}\pi_i(\sigma(\hat{f}_{-i},h_i)(k)) + \frac{1}{t}(t-t_0)\pi_i(\sigma) \leq \pi_i(\sigma) + \frac{1}{\sqrt{t}}$$

for sufficiently large t. If there exists $t_0 \leq t_1$ such that i deserves punishment after $\{\sigma(\hat{f}_{-i},h_i)(k)\}_{k=1}^{t_1}$, then

$$\frac{1}{t}\sum_{k=1}^{t}\pi_i(\sigma(\hat{f}_{-i},h_i)(k)) = \frac{1}{t}\sum_{k=1}^{t_1}\pi_i(\sigma(\hat{f}_{-i},h_i)(k)) + \frac{1}{t}(t-t_1)\pi_i(\sigma) \leq \pi_i(\sigma) + \frac{1}{\sqrt{t}}$$

for sufficiently large t. ∎

5. References

[1] R. J. Aumann, "Acceptable points in general cooperative n-person games," in: A. W. Tucker and R. C. Luce, Eds., *Contributions to the Theory of Games* **IV**, Annals of Mathematics Studies No. 40, Princeton University Press, Princeton, N.J., 1959, pp. 287–324.

[2] R. J. Aumann, "Acceptable points in games of perfect information," *Pacific J. Math.* **10** (1960) 381–417.

[3] R. J. Aumann, "The core of a cooperative game without side payments," *Trans. Amer. Math. Soc.* **98** (1961) 539–552.

[4] R. J. Aumann, "A survey of cooperative games without side payments," in: M. Shubik, Ed., *Essays in Mathematical Economics in Honor of Oskar Morgenstern*, Princeton University Press, Princeton, N.J., 1967, pp. 3–27.

[5] R. J. Aumann, *Lectures on Game Theory*, Stanford University, 1976.

[6] J. W. Friedman,"A non-cooperative equilibrium of supergames," *International Economic Review* **12** (1971) 1–12.

[7] M. Kurz, "Altruistic Equilibrium," Technical Report No. 156, Institute for Mathematical Studies in the Social Sciences, Stanford University, 1975.

[8] R. P. Luce and H. Raiffa, *Games and Decisions*, New John Wiley, New York, 1957.

[9] A. E. Roth, "Self supporting equilibrium in the supergame," unpublished manuscript, 1975.

Essay 3

Entry, Divorce and Re-Marriage in Matching Markets

Elaine Bennett[1]

ABSTRACT. This paper investigates the effect of entry on marriage markets. The paper presents a natural adjustment process for marriage markets and shows that entry on one side of the market makes that side worse off and the other side better off than before entry occurred. An example shows that the adjustment process described by Roth and Vande Vate lacks this property; there may be winners and losers on both sides.

1. Introduction

The recent literature on matching models has yielded valuable insights into the nature of markets for heterogeneous goods. Such models have been used by Gale and Shapley [7] and Gale and Sotomayor [8] to study the matching of men and women (the marriage market), by Crawford and Knoer [3], Kelso and Crawford [9] and Roth [13] to study the matching of firms and workers (the job market), and by Roth [14, 15] to study the the matching of residents and hospitals (e.g., the National Residents Matching program). (See Roth and Sotomayor [16] for further references to the substantial literature.) In this paper we consider one-to-one matchings with strict preferences (marriage markets) and investigate the effect of entry.

An equilibrium in the marriage market (i.e., a *stable matching*) is a specification of married couples-and single men and women such that no married person would prefer to be single and no man and woman would prefer marriage to each other rather than their current situation (either married or single). Gale and Shapley [7] prove the existence of stable matchings by providing a decentralized mechanism which always converges to a stable

[1]Department of Economics, Virginia Polytechnic Institute and State University, Blacksburg VA 24061-0316.

matching. In their "deferred acceptance" procedure, men make proposals to their most preferred women, and women tentatively accept their most preferred proposal and reject the rest. Rejected men then propose to their "next-most-preferred" women, and women again tentatively accept their most preferred proposal and reject the rest. After a finite number of stages, no man makes a new proposal; at this point, women are married to their "tentatively accepted" men. Gale and Shapley show that the resulting matching is stable. Moreover, among all stable matchings, it is the unique stable matching which is unanimously preferred by all men (and unanimously dispreferred by all women) to all other stable matchings; *i.e.*, the *man-optimal* stable matching.

The deferred acceptance procedure displays a certain bias, since (when men make all the proposals) it always converges to man-optimal stable matching.[2] Roth and Vande Vate [17] describe a procedure which lacks this bias. The procedure is based on divorce and remarriage of "blocking" pairs. A *blocking pair* is a woman/man pair who would prefer marriage to each other rather than their current situation (either married to their current spouses or single)[3] At each step of the adjustment process, the blocking pair procedure takes the current matching and produces a new matching by divorcing a blocking pair from their current spouses (if any) and marrying them to each other; their former spouses become single. Roth and Vande Vate prove that, beginning at *any* matching, there is a sequence of divorces and remarriages by blocking pairs that leads to a stable matching; moreover, beginning at the degenerate matching in which all men and women are single, *every* stable matching can be reached by a sequence of divorces and remarriages of blocking pairs. If blocking pairs are chosen at random (with probabilities bounded away from zero) then the procedure converges to a stable matching with probability one and to every stable matching with positive probability.[4]

A number of authors have studied the effect of entry in marriage markets and more general contexts. Kelso and Crawford [9] and Demange and Gale [5] compare the *optimal* stable matchings before and after entry. They show that, after the entry of a woman into the market, the *man-optimal* stable matching for the expanded market is weakly better for all men and weakly worse for all women (in the original market) than the *man-optimal* stable matching in the original market. Similarly the *woman-optimal* stable matching in the expanded market is weakly better for all men and weakly worse for all women (in the original market) than the *woman-optimal* stable

[2]If women propose, the result is the woman-optimal stable matching.

[3]An individual who would prefer being single to his/her current situation is also considered a "blocking pair".

[4]If blocking pairs are not chosen at random, the blocking pair procedure can cycle and never lead to a stable matching. For an example see Knuth [10].

matching of the original market. Roth and Sotomayor [16] (following Mo [11]) compare all stable matchings before and after entry. They show that there is a non-empty set of women who are worse off (and a non-empty set of men who are better off) at *every* stable matching of the expanded market than at *any* stable matching of the original market.

Although these results are quite striking, they are not entirely satisfying, because they do not address a basic question: when a marriage market adjusts to an entrant (say, a woman) can one unambiguously say that *all* men will be made (weakly) better off and *all* women will be made (weakly) worse off than before? Comparisons involving only man- (or woman-) *optimal* stable matchings do not address this question because the outcome may not be the man (or woman) optimal stable matching — even when the initial matching is man- (or woman-) optimal. Comparisons involving all stable matchings have nothing to say about *all* men or *all* women; consistent with these results is the possibility that some woman may be better off and some man worse off than before.

In this paper we address this question by describing a natural decentralized procedure by which a marriage market may adjust to entry. Given a stable matching and an entrant, a woman, say, consider the set of men who prefer the entrant to their current situation (either married or single). Among these, there will be a unique man whom the entrant prefers. If the entrant prefers this man to remaining single, we assume that this man divorces his present spouse and marries the entrant. The newly divorced woman is then in the role of an entrant. Consider the set of men who prefer her to their current spouse; among these, there will be a unique man whom the entrant most prefers; if the entrant prefers this man to remaining single, this man divorces his present spouse and marries the entrant, and so forth. The process stops when either the current entrant prefers to remain single or else her chosen spouse is not married. Since, at each stage, the man and woman who marry prefer each other to every other possibility, we call this procedure *optimal divorce and remarriage*. We show that, beginning with an arbitrary stable matching and a woman entrant, this procedure converges in a finite number of steps to a stable matching having the property that all men are weakly better off and all women (in the original market) are weakly worse off than at the original matching.

The optimal divorce and remarriage procedure is a special case of the Roth and Vande Vate blocking pair procedure, so it is natural to ask whether the latter (more general) procedure shares the property of always leading to a stable matching which is worse for women and better for men. We show by example that it does not.

Following this Introduction, Section 2 presents the optimal divorce and remarriage procedure, and establishes our comparison. Section 3 provides an example comparing it to the blocking pair procedure. Finally, Section 4 provides a discussion.

2. Divorce and Remarriage

The agents in the marriage market consist of a finite (non-empty) set W of women and a finite (non-empty) set M of men. Each of the agents has a strict transitive preference relation over agents of the opposite sex (as a marriage partner) and the possibility of remaining unmarried. It is convenient to represent the latter possibility by matching an agent with herself (or himself). We write $P(w)$ for the strict preference ordering of woman w, $P(m)$ for the strict preference order if man m, and $R(w)$ and $R(m)$ for the associated weak preference orderings. Write P for the list of strict preference orderings of all market participants. Thus the data of a marriage market is a triple (W, M, P).

A *matching* is a function $\mu : M \cup W \to M \cup W$ such that (i) $\mu(\mu(x)) = x$ for each $x \in M \cup W$ (so that if x is married to y then y is married to x), and (ii) if $w \in W$ and $\mu(w) \neq w$ then $\mu(w) \in M$, and if $m \in M$ and $\mu(m) \neq m$ then $\mu(m) \in W$ (so that every agent is either unmarried or married to an agent of the opposite sex). A matching μ is *stable* if there is no married agent who would prefer to be single, and no pair of agents who would prefer to be married to each other rather than to their current spouses (or to remaining single). As Gale and Shapley [7] show, the set of stable matchings is always non-empty.

The preferences of agents of each sex induce partial orderings on the set of stable matchings: If μ and μ' are stable matchings, then $\mu \succeq_w \mu'$ if $\mu(w)R(w)\mu'(w)$ for all $w \in W$ (*i.e.*, all women weakly prefer their mates in the matching μ to their mates in the matching μ'); $\mu' \succeq_m \mu$ if $\mu'(m)R(m)\mu(m)$ for all $m \in M$ (*i.e.*, all men prefer their mates in the matching μ' to their mates in the matching μ). These partial orderings are opposite: $\mu \succeq_w \mu'$ if and only if $\mu' \succeq_m \mu$. Moreover, each of these partial orderings makes the set of stable matchings into a lattice. In particular, there is a unique stable matching μ_w (the *woman-optimal* stable matching) which is unanimously preferred by all women (and hence dispreferred by all men) to every other stable matching, and a unique stable matching μ_m (the *man-optimal* stable matching) which is unanimously preferred by all men (and hence unanimously dispreferred by all women) to every other stable matching.

The situation of interest to us is as follows: Let (W, M, P) and (W', M, P') be marriage markets, such that $W' = W \cup \{w_0\}$, and such that P' agrees with P on W and M. We refer to (W, M, P) as the "original" market and (W', M, P') as the "final" or "expanded" market. Let μ be a stable matching for the market (W, M, P). When woman w_0 enters the market, we may imagine a process of divorces and remarriages. It is convenient (for the inductive description that follows) to write $W_0 = W$ and $\mu_0 = \mu$. Let $G(w_0, \mu_0)$ be the set of men who prefer marriage to woman w_0 to their state at the matching μ_0. If $G(w_0, \mu_0)$ is not empty, let m_1 be the man who ranks highest in the (strict) preference ordering of woman w_0. If w_0 prefers m_1 to remaining single, set $w_1 = \mu_0(m_1)$, and let μ_1 be the matching of $W_1 = W_0 \cup \{w_0\} \setminus \{w_1\}$ with M which replaces w_1 with w_0 and otherwise agrees with μ_0. We can now think of w_1 as a entrant in this market, and repeat the process. The process terminates if the entrant w_k marries a single man or if w_k prefers to remain single rather than marry any man in $G(w_k, \mu_k)$. Since the entrant always marries her first choice (among men who are willing to marry her rather than maintain their current state), and her new spouse marries his first choice (among women who are willing to marry him rather than maintain their current state), we refer to the pair as an *optimal blocking pair* and refer to this process as *optimal divorce and remarriage.*

The following result provides the comparison we seek.

Theorem 2.1.

(i) The optimal divorce and remarriage procedure terminates after a finite number of steps.

(ii) The terminal matching μ^* is stable.

(iii) All men (weakly) prefer μ^* to μ; all women in W (weakly) prefer μ to μ^*.[5]

Proof: We show first that the restriction of the matching μ_1 to $W_1 \cup M$ is stable. To see this, suppose (to the contrary) that the woman w and the man m are *not* married to each other at the matching μ_1 and would prefer each other to their respective mates $\mu_1(w)$, $\mu_1(m)$ at μ_1. If $w \neq w_0$ and $m \neq \mu_1(w_0)$, then $\mu_1(w) = \mu_0(w)$ and $\mu_1(m) = \mu_0(m)$; but this implies that μ_0 is also not stable, a contradiction. If $w = w_0$ and $m \neq \mu_1(w_0)$, we obtain a contradiction because $\mu_1(m) = \mu_0(m)$ and the entrant w_0 married a man who preferred her to his previous mate. If $m = \mu_1(w_0)$ and $w \neq w_0$, we obtain a contradiction because $\mu_1(w) = \mu_0(w)$ and the new mate of the entrant w_0 preferred her to all women who were willing to marry him. We conclude that the restriction of μ_1 to $W \cup M$ is a stable matching, as asserted.

It now follows by the obvious inductive argument that each of the matchings $\mu_0, \mu_1, \mu_2, \ldots$ is stable, and hence that the terminal matching μ^* is stable (assuming of course that the process terminates.)

[5] Of course, woman w_0 weakly prefers μ^*, since she was unmarried at μ.

To see that men prefer μ^* to μ, consider the k-the stage: the man who is displaced at the matching μ_{k+1} improves his position, while all other men retain the same spouses. Hence all men (weakly) prefer μ_{k+1} to μ_k. Transitivity of preferences thus implies that all men (weakly) prefer μ^* to μ; indeed, any man who has been divorced at *any* stage strictly prefers μ^* to μ.

To see that women prefer μ to μ^*, consider the k-th stage. The woman w who is displaced at the matching μ_k (and is single in the matching μ_{k+1} either remains single or remarries at stage $k+2$. If she remains single, then the process terminates, and she is strictly worse off (since μ_k is stable, w prefers her old spouse $\mu_k(w)$ to remaining single). On the other hand, if w remarries at stage $k+2$, then her new spouse $\mu_{k+2}(w)$ is inferior to her old spouse $\mu_k(w)$ (since he was available but not chosen). In either case, we conclude that w strictly prefers μ_k to μ_{k+2}. Transitivity of preferences implies that all women who are ever divorced strictly prefer μ to μ^*. Of course, women who are never divorced are indifferent between μ and μ^* .

Since some man is made strictly better off at each stage (and never becomes worse off), this process must terminate after a finite number of stages. This completes the proof. ∎

It is not hard to show that, if the initial matching μ is the woman-optimal matching for (W, M, P), then the final matching μ^* is the woman-optimal matching for (W', M, P').[6] If the initial matching μ is the man-optimal matching for (W, M, P), then the final matching μ^* need not be the man-optimal matching for (W', M, P'); however, since the man-optimal matching μ'_M is better for all men (and worse for all women) than μ^*, it follows that μ'_M is better for all men (and worse for all women) than μ_M. Hence we obtain an alternate proof of the following result.

Corollary 2.2.

(i) *All men (weakly) prefer the woman-optimal stable matching μ'_W to the woman-optimal stable matching μ_W, and all women in W weakly prefer μ to μ'_W.*

(ii) *All men (weakly) prefer the man-optimal stable matching μ'_M to the man-optimal stable matching μ_M, and all women in W (weakly) prefer μ_M to μ'_M.*

---

[6]The easiest way to see this is to recall that the woman-optimal matching can be obtained by the Gale-Shapley deferred acceptance procedure, and then to note that the procedure given here reduces, in that case, to simply continuing the deferred acceptance procedure for one additional woman. We thank Al Roth for pointing this out.

3. An example

In this Section we give a simple example to show that entry of a woman, followed by divorces and remarriages of arbitrary blocking pairs, may lead to a stable matching that is not worse for the women (or better for the men). The initial marriage market contains two women: w_1 and w_2 and three men: m_1, m_2 and m_3. The entrant is woman w_0. The preferences of market participants are given below; preferences before entry are obtained by deleting woman w_0 and her preferences.

$$\begin{array}{lll} P(w_0) = m_2,\ m_3,\ m_1,\ w_0 & & P(m_1) = w_0,\ w_2,\ w_1,\ m_1 \\ P(w_1) = m_1,\ m_2,\ m_3,\ w_1 & & P(m_2) = w_0,\ w_1,\ w_2,\ m_2 \\ P(w_2) = m_2,\ m_1,\ m_3,\ w_2 & & P(m_3) = w_1,\ w_2,\ w_0,\ m_3 \end{array}$$

The man-optimal stable matching μ_M for the original market (the market without w_0) is:

$$w_1 \leftrightarrow m_2 \qquad w_2 \leftrightarrow m_1 \qquad m_3 \leftrightarrow m_3 \ .$$

That is, w_1 and m_2 are married, and m_3 is single.

Using optimal blocking pairs : If w_0 enters, all three men prefer marriage to her rather than their current state. If w_0's favorite – m_2 – divorces his mate – w_1 – and marries w_0, the resulting stable matching (for the market without w_1 is:

$$w_0 \leftrightarrow m_2 \qquad w_2 \leftrightarrow m_1 \qquad m_3 \leftrightarrow m_3 \ .$$

When w_1 re-enters, only m_3 prefer hers to his current state; the resulting stable matching is:

$$w_0 \leftrightarrow m_2 \qquad w_2 \leftrightarrow m_1 \qquad w_1 \leftrightarrow m_3 \ .$$

Here the process terminates. This matching is the man-optimal stable matching for the expanded market. As asserted by Theorem 2.1 comparing this matching with the initial matching μ_M, we find that men m_2 and m_3 are strictly better off, and m_1 is indifferent while w_1 is strictly worse off and w_2 is indifferent.

Using arbitrary blocking pairs: When woman w_0 enters the original market (at the matching μ_M) w_0 / m_1 is a blocking pair (but *not* an optimal blocking pair since w_0 prefers m_2), if they marry, the resulting matching (for the market without woman w_2) is:

$$w_1 \leftrightarrow m_2 \qquad w_0 \leftrightarrow m_1 \qquad m_3 \leftrightarrow m_3 \ .$$

When w_2 re-enters, w_2 / m_2 is a blocking pair, their marriage leads to:

$$w_1 \leftrightarrow m_2 \qquad w_0 \leftrightarrow m_1 \qquad w_2 \leftrightarrow m_3 \ .$$

This matching is not a stable because w_0 / m_2 is a blocking pair; if they divorce their spouses and marry each other we obtain:

$$w_1 \leftrightarrow w_1 \qquad w_0 \leftrightarrow m_2 \qquad w_2 \leftrightarrow m_3 \qquad m_1 \leftrightarrow m_1 \ .$$

This matching is also not stable because w_1 / m_1 is a blocking pair; their marriage leads to:

$$w_0 \leftrightarrow m_2 \qquad w_2 \leftrightarrow m_3 \qquad w_1 \leftrightarrow m_1 \ .$$

There are no blocking pairs for this matching, so the process terminates. This matching is the woman-optimal stable matching for the expanded market. In this matching w_1 and m_2 are both strictly better off while w_2 and m_1 are strictly worse off than in the original matching μ_M. As this example makes clear, if divorce and remarriage take place for *arbitrary* blocking rather than *optimal* blocking pairs, situations can arise where entry leads to winners and losers on both sides of the marriage market.

4. Concluding remarks

Incentives: In the example of Section 3, w_0 choosing her optimal spouse in the first round leads to the man-optimal stable matching in the expanded market, while w_0 choosing a non-optimal spouse in the first round leads to the woman-optimal stable matching. Of course all women (weakly) prefer the latter outcome, but in this case at least, w_0 is indifferent between the two (since she has the same mate in either case). This suggests an interesting question about incentives: Is it possible (in some situations) for some farsighted woman, once divorced, to marry a man *not* her favorite, and thereby manipulate the final outcome in her own favor?[7]

Exit: Although we have analyzed the case of entry by a woman, our results apply equally well for the case of entry by a man: after the market adjusts (assuming *optimal* blocking pairs marry) to the entry of a man, all men will be made worse off and all women will be made better off. Less obvious is that our analysis also applies to *exit* . To see this, consider a stable matching, and suppose that woman w, currently married to man m, *leaves* the market. We may view this situation as entry by man m into the stable matching obtained by deleting the married couple (w, m). After the market adjusts (assuming optimal blocking pairs marry) to the exit of woman w, all remaining women will be made better off and all men will be made worse off.

---

[7]The literature on the strategic aspects of the deferred acceptance procedure suggests that the answer might be "no"; see Chapter 4 in Roth and Sotomayor, as well as Dubins and Freedman [6] and Gale and Sotomayor [8].

Bargaining: "Marriage" markets with strict preferences are appropriate for modeling situations where the negotiable aspects of each partnership either *cannot* change the preference order (e.g., "no amount of money could get me to give up a job at A for a job at B") or where all negotiations have already taken place. When preferences over potential partners may depend on "yet to be" negotiated details of the partnership, a bargaining model of the marriage market may be more appropriate. See, for instance Bennett [1], Crawford and Rochford [4] or Rochford [12].

Acknowledgments: Financial support from the National Science Foundation Grant SES 87-06631 is gratefully acknowledged.

5. References

[1] E. Bennett, "Consistent Bargaining Conjectures in Marriage and Matching", *Journal of Economic Theory* **45** (1988) 392–407.

[2] V. Crawford, "Comparative Statics in Matching Markets," *Journal of Economic Theory*, to appear.

[3] V. Crawford and E. Knoer (1981), "Job Matching with Heterogeneous Firms and Workers," *Econometrica* **49** (1981) 437–450.

[4] V. Crawford and S. Rochford, "Bargaining and Competition in Matching Markets," *International Economic Review* **49** (1986) 329–348.

[5] G. Demange and D. Gale, "The Strategy Structure of Two-sided Matching Markets," *Econometrica* **53** (1986) 873–888.

[6] L. Dubins and D. Freedman, "Machiavelli and the Gale-Shapley Algorithm," *American Mathematical Monthly* **88** (1981) 485–494.

[7] D. Gale and L. S. Shapley, "College Admissions and the Stability of Marriage," *American Mathematical Monthly* **69** (1962) 9–15.

[8] D. Gale and M. Sotomayor (1985), "Ms. Machiavelli and the Stable Matching Problem," *American Mathematical Monthly* **92** (1985) 261–268.

[9] A. Kelso and V. Crawford, "Job Matching, Coalition Formation, and Gross Substitutes," *Econometrica* **50** (1982) 1483–1504.

[10] D. E. Knuth, *Marriages Stables*, Les Presses de l'Universite de Montreal, 1976.

[11] J. Mo, "Entry and Structures of Interest Groups in Assignment Games," *Journal of Economic Theory* **46** (1978) 66–96.

[12] S. Rochford, "Symmetrically Pairwise-bargained Allocations in an Assignment Market," *Journal of Economic Theory* **34** (1984) 262–281.

[13] A. E. Roth, "Stability and Polarization of Interests in Job Matching," *Econometrica* **52** (1984) 47–57.

[14] A. E. Roth, "The Evolution of the Labor Market for Medical Interns and Residents: A Case Study in Game Theory," *Journal of Political Economy* **36** (1984) 277–288.

[15] A. E. Roth, "New Physicians: A Natural Experiment in Market Organization," *Science* **250** (1990) 1524–1528.

[16] A. E. Roth and M. Sotomayor, *Two-Sided Matching: A Study in Game-Theoretic Modelling and Analysis*, Cambridge University Press, 1990.

[17] A. E. Roth and J. Vande Vate, "Random Paths to Stability in Two-Sided Matching," *Econometrica* **58** (1990) 1475–1480.

[18] L. S. Shapley and M. Shubik (1972), "The Assignment Game I: the Core," *International Journal of Game Theory* **1** (1972) 23–28.

Essay 4

Games in Support Function Form: An Approach to the Kernel of NTU Games

Louis J. Billera[1] **and Richard P. McLean**[2]

1. Introduction

In this paper, we extend the kernel for transferable utility (TU) games to the nontransferable utility (NTU) case using methods of convex analysis. The kernel for TU games was introduced by Davis and Maschler [9] and its mathematical structure was further developed by Maschler and Peleg [15], [16]. The kernel is a subset of the bargaining set (Aumann and Maschler [3]) that contains the nucleolus (Schmeidler [23]). For more results and references, see Maschler, Peleg, and Shapley [17], [18].

As with any solution concept in cooperative game theory, it is desirable to extend the results from the transferable utility case to the nontransferable utility setup and such extensions have been successfully developed for the core and the Shapley value (Shapley [25]). The bargaining set of an NTU game has been explored by Peleg [21], Billera [5], and Asscher [4] while approaches to the kernel and nucleolus of an NTU game were offered by Kalai [11] and Nakayama [19].

In the present work, which represents a completion of the unpublished report by Billera [7], we consider an alternative approach to the kernel of an NTU game using methods of convex analysis. In particular, we represent a game not as a collection of "attainable sets" but rather as a collection of "support functions". This will enable us to prove an existence theorem using the machinery of Billera [5]. The kernel that we define has several aesthetically desirable properties. For a large class of NTU games, the kernel is nonempty and for NTU games derived from TU games, our kernel coincides with the previously defined kernel of a TU game.

---

[1]Department of Mathematics and School of Operations Research and Industrial Engineering, Cornell University, Ithaca, NY 14853

[2]Department of Economics and RUTCOR, Rutgers University, New Brunswick, NJ 08903

The paper is structured as follows. In section 1, we record the relevant concepts from convex analysis and game theory. In section 2, we introduce games in support function form and discuss their relationship with TU and NTU games. Section 3 contains the definition of the kernel of a game in support function form as well as the main existence theorem. In section 4, we discuss the relationship between the kernel and some other solution concepts for NTU games.

2. Notation and Definitions

Throughout the paper, N denotes the finite set $\{1, \ldots, n\}$ and 2^N denotes the power set of N. $\Re^n_+$ ($\Re^n_{++}$) represents the nonnegative (positive) orthant of $\Re^n$. For $S \in 2^N$, define the subspace $\Re^S = \{x \in \Re^n | x_i = 0 \text{ for } i \notin S\}$ and $\Re^S_+ = \Re^n_+ \cap \Re^S$. For $x \in \Re^n$, let x^S denote its projection into $\Re^S$. The vector $(1, 1, \ldots, 1)$ in $\Re^n$ is denoted χ. Therefore, $\chi^S_i = 1$ if $i \in S$ and 0 if $i \notin S$. If λ, $x \in \Re^n$, then $\lambda * x \in \Re^n$ is the vector whose ith component is $(\lambda * x)_i = \lambda_i x_i$ while $\lambda \cdot x$ denotes the inner product.

All concepts from convex analysis can be found in Rockafellar [22] or Aubin and Ekeland [1]. If $K \subseteq \Re^n$ is a cone, define the (negative) polar cone as $K^* = \{y \in \Re^n | y \cdot x \leq 0 \text{ for all } x \in K\}$. For a nonempty, convex set $C \subseteq \Re^n$, define the recession cone of C as $0^+C = \{y \in \Re^n | C + y \subseteq C\}$ and the barrier cone of C as $b(C) = \{y \in \Re^n | \sup_{x \in C} x \cdot y < \infty\}$. The cones 0^+C and $b(C)$ are convex cones containing the origin and $b(C)^* = 0^+C$. The recession cone consists of those directions in which C is unbounded.

For a closed, convex set C, define the extended real valued support function of C (denoted σ_C) as $\sigma_C(\alpha) = \sup_{x \in C} \alpha \cdot x$. The set C can be recovered from its support function: in particular, $C = \{x | \alpha \cdot x \leq \sigma_C(\alpha), \text{ for all } \alpha\}$. The set $\{\alpha | \sigma_C(\alpha) < \infty\}$ is the effective domain of σ_C (denoted $\operatorname{dom} \sigma_C$) and therefore, $b(C) = \operatorname{dom} \sigma_C$. The support function σ_C of a nonempty closed convex set C is convex, lower semicontinuous, positively homogeneous and proper (i.e. $\sigma_C(\alpha) > -\infty$ for all α and $\sigma_C(\alpha) < \infty$ for some α). In fact, any convex, lower semicontinuous, positively homogeneous, proper extended real valued function f is the support function of the nonempty closed convex set defined as $\{x | \alpha \cdot x \leq f(\alpha), \text{ for all } \alpha\}$.

A *transferable utility game* in coalitional form (TU game for short) is a set function $v : 2^N \to R$ such that $v(\emptyset) = 0$. A TU game is zero normalized if $v(i) = 0$ for all $i \in N$. The set of zero normalized TU games that also satisfy $v(N) \geq 0$ is denoted G.

A *nontransferable utility game* in coalitional form (NTU game for short) is a correspondence that associates with each nonempty $S \subseteq N$ a set $V(S)$ satisfying

1 $V(S)$ is a non-empty, closed, convex subset of $\Re^S$,

2 $b(V(S))$ is a convex polyhedral cone such that $b(V(S)) \cap \Re^S \subseteq \Re^S_+$ and

3 $\bigcap_{S \subseteq N} b(V(S)) \neq \{0\}$.

An NTU game is zero normalized if $V(\{i\}) = \{x \in \Re^{\{i\}} | x_i \leq 0\}$, for all $i \in N$. Let Γ denote the class of zero normalized NTU games such that $V(N) \cap \Re^n_+$ is nonempty and compact.

Condition (i) is standard in many treatments of NTU games (*e.g.*, Aumann [2], Hart [10]). Condition (ii) implies that $-\Re^S_+ \subseteq 0^+(V(S))$ and therefore each $V(S)$ is comprehensive ($x \in V(S)$ whenever $y \in V(S)$ and $x \leq y$). The polyhedral restriction is needed to insure that the support function of each $V(S)$ is continuous on its effective domain. Related conditions appear in Kern [14] where the continuity of the support functions is required to prove the existence of Shapley NTU values. Condition (iii) says that there is at least one nonzero α such that $\sigma_{V(S)}(\alpha) < \infty$ for all $S \subseteq N$.

Γ contains all zero normalized NTU games where $V(N) \cap \Re^n_+$ is nonempty and compact and each $V(S)$ is the (Minkowski) sum of a convex, compact subset C_S of $\Re^S$ and a convex polyhedral cone K_S where $-\Re^S_+ \subseteq K_S \subseteq \Re^S$ (in this case $0^+V(S) = K_S$). This is a result of the easily proven fact that, if $A = C + K$ is the sum of a compact set C and a closed, convex cone K, then $b(A) = K^*$. In addition, the space G can be embedded in Γ in the usual manner: if v is a TU game, define $V(S) = \{x \in \Re^S | \chi^S \cdot x \leq v(S)\}$. In this case, $-\Re^S_+ \subseteq 0^+V(S) = \{x \in \Re^S | \chi^S \cdot x \leq 0\}$ and $b(V(S)) = \{x \in \Re^n | x^S = t\chi^S$ for some $t \geq 0\}$.

3. Games in Support Function Form

For an NTU game $V \in \Gamma$ and a set $S \subseteq N$, the support function $\sigma_{V(S)}$ of the attainable set $V(S)$ has the following properties. Each $\sigma_{V(S)}$ is a proper, convex, positively homogeneous, lower semicontinuous extended real valued function such that $\sigma_{V(S)}(\alpha) = \sigma_{V(S)}(\alpha^S)$ for all α. In addition $\sigma_{V(S)}$ is a (finite) continuous function on $b(V(S))$. This follows since support functions are always lower semi-continuous [22] (Theorem 13.2), $b(V(S))$ is a convex polyhedral cone and a convex function is upper semi-continuous relative to any polyhedral set on which it is finite [22] (Theorem 10.2).

Given $\sigma_{V(S)}$, one can recover $V(S)$ in the manner described above *i.e.*,

$$V(S) = \{x \in \Re^S \mid x \cdot \alpha \leq \sigma_{V(S)}(\alpha) \text{ for all } \alpha \in \Re^n\}.$$

However, the various properties of $\sigma_{V(S)}$ enumerated above can be used to construct a "dual" definition of a nontransferable utility game that will be useful in extending the notion of kernel to NTU games.

Definition 3.1. An NTU game in support function form is a collection $f = \{f_S\}_{\emptyset \neq S \subseteq N}$ of convex, proper, positively homogeneous extended real valued functions on $\Re^n$ such that

1 $f_S(\alpha) = f_S(\alpha^S)$ for all $\alpha \in \Re^n$

2 $\mathrm{dom}\, f_S$ is a closed subset of $\Re^S_+ + \Re^{N\setminus S}$ on which f_S is continuous

3 $D_f := \bigcap_{S \subseteq N} \mathrm{dom}\, f_S \neq \{0\}$.

A game in support function form can be given the following interpretation. Each $\alpha \in \mathrm{dom}\, f_S$ represents a weighting of the players' utilities for which utility becomes "transferable" and $f_S(\alpha)$ is the maximum total utility attainable for these weights. That is, a payoff vector x is attainable by S for the transfer weights α if $\alpha \cdot x \leq f_S(\alpha)$. In actuality, the only payoffs attainable by S are those that are attainable for every transfer weight vector $\alpha \in \mathrm{dom}\, f_S$ which is precisely the set

$$V_f(S) = \{x \in \Re^n | \alpha \cdot x \leq f_S(\alpha) \text{ for all } \alpha \in \Re^n\}.$$

Note that condition (i) of Definition 3.1 implies that $V_f(S) \subseteq \Re^S$. Denote by $\mathcal{F}$ the class of all SFF games for which $f_{\{i\}} \equiv 0$ and $V_f(N) \cap \Re^n_+$ is nonempty and compact.

To analyze the relationship between an NTU game in Γ and a game in support function form, suppose that $f \in \mathcal{F}$ and $S \subseteq N$, and let $V_f(S)$ be the set defined above. If for each S, $\mathrm{dom}\, f_S$ is a convex polyhedral cone, then $V_f \in \Gamma$. Conversely, if $V \in \Gamma$, then the collection of support functions $\{\sigma_{V(S)}\}$ is a game in support function form.

If $V \in \Gamma$ is derived from the TU game v, then the collection of support functions $\{\sigma_{V(S)}\}$ is an SFF game such that for each S, $\sigma_{V(S)}(\chi) = v(S)$ and $\mathrm{dom}\, \sigma_{V(S)} = \{x \in \Re^n | x^S = t\chi^S \text{ for some } t \geq 0\}$. Conversely, let $\{f_S\}$ be an SFF game such that for each S, $\mathrm{dom}\, f_S = \{x \in \Re^n | x^S = t\chi^S \text{ for some } t \geq 0\}$. Then $\{f_S\}$ is the game in support function form derived from the TU game v where $v(S) = f_S(\chi)$.

As a consequence of the discussion in the preceding paragraphs, we will call an SFF game an NTU (resp. a TU game) game if $\mathrm{dom}\, f_S$ is a convex polyhedral cone (resp. $\mathrm{dom}\, f_S = \{x \in \Re^n | x^S = t\chi^S \text{ for some } t \geq 0\}$) for each S.

The support function approach was used by Billera [6] to give a characterization of those NTU games with a non-empty core. This type of description has also been used by Baudier [8].

4. The Kernel

Let v be a TU game on N and for each $x \in \Re^n$, define the *excess* of S at x, denoted $e_S(x)$, as

$$e_S(x) = v(S) - \chi^S \cdot x.$$

If $x \in \Re^n$ and $i, j \in N$, define the *surplus* of i over j at x, denoted $s_{ij}(x)$, as

$$s_{ij}(x) = \max\{e_S(x) \mid i \in S, j \notin S\},$$

If $v \in G$, define the set of imputations as:

$$X(v) = \{x \in \Re^n_+ \mid \chi \cdot x = v(N)\}.$$

Definition 4.1. Let $v \in G$. The *kernel* of v, denoted ker(v)), is the set

$$\{x \in X(v) \mid \text{ for all } i, j \in N, \text{ either } s_{ij}(x) \leq s_{ji}(x) \text{ or } x_j = 0\}\ .$$

Remark 4.2. We have defined the kernel for the "grand coalition." One can of course define the kernel for each partition of N (i.e. for each "coalition structure") as in Davis and Maschler [9]. All of our results can be extended to the case of arbitrary coalition structures with appropriate modifications of the definitions. The nonemptiness of the kernel of a game $v \in G$ is proved in [9].

Several approaches to the kernel of an NTU game are possible. First, one could use "generalized excess functions" as presented by Kalai [11]. In that work, one defines a collection of real valued functions $\{h_S\}_{S \subseteq N}$ that share certain abstract features with the usual excess functions defined for TU games (e.g. $h_S(x) \geq 0$ if $x \in V(S)$, $h_S(x) = 0$ if x belongs to the Pareto boundary of $V(S)$, etc). A "natural" example that coincides with the usual excess for TU games is given by $h_S(x) = \sup\{t | x^S + (t/|S|)\chi^S \in V(S)\}$. Given a collection of excess functions, the surplus functions and the kernel can be defined for that collection and therefore, different kernels can be constructed for different collections of excess functions.

An alternative "λ-transfer" approach that imitates the methodology of the Shapley NTU value is also possible (Shapley [25], Aumann [2]). For $\lambda \in \bigcap_{S \subseteq N} \operatorname{dom} \sigma_{V(S)}$, define a TU game v_λ as $v_\lambda(S) = \sigma_{V(S)}(\lambda)$. Next extend the notion of imputation to NTU games by defining:

$$X(V) = \{x \in V(N) \cap \Re^n_+ \mid \text{ there is no } y \in V(N) \text{ such that } y > x\}.$$

Finally, define the NTU *kernel* of $V \in \Gamma$ as the set

$$\text{NTU-ker}(V) = \{x \in X(V) \mid \lambda{*}x \in \ker(v_\lambda) \text{ for some } 0 \neq \lambda \in \bigcap_S \operatorname{dom} \sigma_{V(S)}\}.$$

Note that the NTU kernel of a game $V \in \Gamma$ derived from a TU game $v \in G$ will coincide with $\ker(v)$ since $0 \neq \lambda \in \bigcap_{S \subseteq N} \operatorname{dom} \sigma_{V(S)}$ if and only if $\lambda = t\chi$ for some $t > 0$ in which case, $v_\lambda = tv$.

We now extend the kernel to NTU games via games in support function form. For an SFF game f, we let $X(f) = X(V_f)$ and extend the notion of excess and surplus to SFF games as follows. For $x \in \Re^n$ and $\alpha \in \operatorname{dom} f_S$, define the *excess* of S at x and α as

$$e_S(x, \alpha) = f_S(\alpha) - \alpha^S \cdot x.$$

Notice that we can consider $e_S(x, \cdot)$ as the restriction to $\operatorname{dom} f_S$ of the support function of the translate of the set $V_f(S)$ by the point $-x^S$. When we consider the excess as a function

$$e_S : \Re^n \times \operatorname{dom} f_S \to \Re,$$

it is clear that it is continuous by part (ii) of Definition 3.1 and positively homogeneous in α.

Finally for $x \in \Re^n$ and $i, j \in N$ and $\alpha \in D_f$, define

$$s_{ij}(x, \alpha) = \max\{e_S(x, \alpha) | i \in S, j \notin S\}.$$

It follows ([22] Corollary 16.5.1) that $s_{ij}(x, \cdot)$ may be considered as the restriction to D_f of the support function of the convex hull of the union of the sets "supported" by the $e_S(x, \cdot)$ involved in the supremum. When we write $s_{ij}(x, \cdot) > s_{ji}(x, \cdot)$, we will mean that $s_{ij}(x, \alpha) > s_{ji}(x, \alpha)$ for each $\alpha \in D_f$ with $\alpha \neq 0$. In terms of the relevant sets, $s_{ij}(x, \cdot) > s_{ji}(x, \cdot)$ means that the set "supported" by $s_{ji}(x, \cdot)$ is contained in the relative interior of the set "supported" by $s_{ij}(x, \cdot)$.

Definition 4.3. Let f be a SFF game. The *kernel* of f, denoted $\ker(f)$, is the set

$$\{x \in X(f) \mid \text{ for all } i, j \in N, \text{ either } \neg(s_{ij}(x, \cdot) > s_{ji}(x, \cdot)) \text{ or } x_j = 0\}$$

where the symbol $\neg$ denotes negation.

This definition has one very desirable property: it is consistent with the definition of kernel given above for TU games. The next result makes this precise.

Proposition 4.4. *If f is the SFF game generated by the TU game v, then* $\ker(f) = \ker(v)$.

Proof: If f is the SFF game generated by $v \in G$, then for each S, $v(S) = f_S(\chi)$ and $\operatorname{dom} f_S = \{x \in \Re^n | x^S = t\chi^S \text{ for some } t \geq 0\}$. Therefore

$D_f = \{x \in \Re^n | x = t\chi \text{ for some } t \geq 0\}$ is the halfline generated by χ. Now let $x \in \Re^n$. We will show that $s_{ij}(x) > s_{ji}(x)$ (in the TU game v) if and only if $s_{ij}(x,\alpha) > s_{ji}(x,\alpha)$ for all $\alpha \in D_f, \alpha \neq 0$. If $\alpha = t\chi$ and $t > 0$, then the homogeneity of $e_S(x,\alpha)$ in α implies that

$$\begin{aligned} s_{ij}(x,\alpha) &= s_{ij}(x,t\chi) \\ &= t \max\{f_S(\chi) - \chi^S \cdot x \mid i \in S, j \notin S\} \\ &= ts_{ij}(x). \end{aligned}$$

Thus, $s_{ij}(x,t\chi) > s_{ji}(x,t\chi)$ for all $t > 0$ if and only if $s_{ij}(x) > s_{ji}(x)$ and since $X(f) = X(v)$, the result follows. ∎

The next result is the main theorem of the paper.

Theorem 4.5. *For every SFF game* $f \in \mathcal{F}$, $\ker(f) \neq \emptyset$.

Proof: For each $x \in X(f)$, define the binary relation $Q(x)$ on N as follows: for all $i,j \in N$, $iQ(x)j$ if and only if $x_j > 0$ and $s_{ij}(x,\cdot) > s_{ji}(x,\cdot)$. We will show that $Q(x)$ is transitive and open (*i.e.*, for all $i,j \in N$, $\{x \in X(f) | iQ(x)j\}$ is open in $X(f)$). It will then follow from an application of [5] (Theorem 1) that $\ker(f) \neq \emptyset$.

To show transitivity, suppose $i,j,k \in N$ and $s_{ij}(x,\cdot) > s_{ji}(x,\cdot)$, $s_{jk}(x,\cdot) > s_{kj}(x,\cdot)$, $x_j > 0$, and $x_k > 0$. We must show that $s_{ik}(x,\cdot) > s_{ki}(x,\cdot)$, i.e. that for each $\alpha \in D_f, \alpha \neq 0$, $s_{ik}(x,\alpha) > s_{ki}(x,\alpha)$. Following the proof of [9] (Lemma 5.1), define for each fixed $\alpha \in D_f, \alpha \neq 0$, the sets

$$\begin{aligned} \mathcal{C} &= \{T \subseteq N \mid 1 \leq |T \cap \{i,j,k\}| \leq 2\}, \text{ and} \\ \mathcal{D} &= \{R \in \mathcal{C} \mid e_R(x,\alpha) \geq e_T(x,\alpha) \text{ for all } T \in \mathcal{C}\}. \end{aligned}$$

It follows that for each $R \in \mathcal{D}$ we have $i \in R$ and $k \notin R$. Hence, for $R \in \mathcal{D}$ and any T containing k and not i,

$$s_{ik}(x,\alpha) \geq e_R(x,\alpha) > e_T(x,\alpha),$$

and so $s_{ik}(x,\alpha) > s_{ki}(x,\alpha)$.

To show that $Q(x)$ is open, suppose $s_{ij}(x,\cdot) > s_{ji}(x,\cdot)$ and $x_j > 0$. Let $H(x,\alpha) = s_{ij}(x,\alpha) - s_{ji}(x,\alpha)$. The function $H : X(f) \times D_f \to \Re$ is continuous and $H(x,\alpha) > 0$ for all $\alpha \in D_f$, $\alpha \neq 0$. Since H is positively homogeneous in α, it is enough to show that there exists an open neighborhood U of x in $X(f)$ such that $H(y,\alpha) > 0$ for all $y \in U$ and all $\alpha \in A = \{\xi \in D_f \mid \|\xi\| = 1\}$.

For each $\alpha \in A$, there exist open neighborhoods $U_\alpha \ni x$ and $V_\alpha \ni \alpha$ (in $X(f)$ and A, respectively) such that $H(y,\beta) > 0$ for all $(y,\beta) \in U_\alpha \times V_\alpha$.

Since A is compact, there exists a collection $V_{\alpha_1}, \ldots, V_{\alpha_k}$ that covers A. Let $U = U_{\alpha_1} \cap \ldots \cap U_{\alpha_k}$ and let $y \in U$. Then for $\alpha \in A$, we have $\alpha \in V_{\alpha_i}$ for some i, $1 \leq i \leq k$. Since $(y, \alpha) \in U_{\alpha_i} \times V_{\alpha_i}$ we have $H(y, \alpha) > 0$.

To complete the proof, define for each $x \in X(f)$ a new binary relation $R(x)$ on N as $iR(x)j$ if and only if $\neg(iQ(x)j)$, *i.e.*, for all $i, j \in N$, $iR(x)j$ if and only if $x_j = 0$ or $\neg(s_{ij}(x, \cdot) > s_{ji}(x, \cdot))$. For $i \in N$, let

$$E_i = \{x \in X(f) \mid jR(x)i \text{ for all } j \in N \setminus \{i\}\}.$$

Note that (a) $x \in X(f)$ and $x_i = 0$ imply that $x \in E_i$; (b) $E_i = X(f) \setminus \{x \in X(f) | jQ(x)i \text{ for some } j \in N \setminus \{i\}\}$ is closed in $X(f)$; (c) for each $x \in X(f)$, there is an $i \in N$ such that $x \in E_i$ because $Q(x)$ is transitive and acyclic. Therefore, by Theorem 1 of [5], $\ker(f) = \bigcap_{i \in N} E_i \neq \emptyset$. ∎

5. Some Properties of the Kernel

The kernel that we have defined can be related to the NTU kernel discussed at the beginning of section 3. Our first result in this direction is:

Proposition 5.1. *Suppose $V \in \Gamma$ and $D_V \setminus \{0\} \subseteq \Re^n_{++}$ where $D_V = \bigcap_{S \subseteq N} \operatorname{dom} \sigma_{V(S)}$. Then* NTU-ker$(V) \subseteq$ ker(V).

Proof: Let V and D_V be as above and suppose that $x \in X(V)$ but $x \notin \ker(V)$. Then there exists i and j such that $x_j > 0$ and for all $0 \neq \alpha \in D_V$, $s_{ij}(x, \alpha) > s_{ji}(x, \alpha)$. Hence, for all $0 \neq \alpha \in D_V$, $\alpha_j x_j > 0$ and $s_{ij}(x * \alpha) > s_{ji}(x * \alpha)$ where $s_{ij}(\cdot)$ denotes the surplus associated with the TU game $v_\alpha(S) = \sigma_{V(S)}(\alpha)$. Therefore, $\alpha * x \notin \ker(v_\alpha)$ so $x \notin$ NTU-ker(V). ∎

For the class of games that we describe next, the kernel and the NTU kernel coincide. An NTU game $V \in \Gamma$ is a hyperplane game if there exists $\beta \in \Re^n_{++}$ and $\gamma \in R$ such that $V(N) = \{x \in \Re^n | \beta \cdot x \leq \gamma\}$. The next result describes the kernel of a hyperplane game.

Proposition 5.2. *Suppose that $V \in \Gamma$ is a hyperplane game with $V(N) = \{v \in \Re^n | \beta \cdot x \leq \gamma\}$. Then* ker$(V) =$ NTU $-$ ker(V).

Proof: For a hyperplane game V, $\bigcap_{S \subseteq N} \operatorname{dom} \sigma_{V(S)} = \{x \in \Re^n | x = t\beta$ for some $t \geq 0\}$. Let $s_{ij}(x, \alpha)$ be the surplus of i over j in V and let $s_{ij}(x)$ denote the surplus of i over j in v_β, where $v_\beta(S) = \sigma_{V(S)}(\beta)$. Since the surplus functions are homogeneous, there exists $\alpha \in \bigcap_{S \subseteq N} \operatorname{dom} \sigma_{V(S)}$ with $\alpha \neq 0$ such that $s_{ij}(x, \alpha) \leq s_{ji}(x, \alpha)$ if and only if $s_{ij}(\beta * x) \leq$

$s_{ji}(\beta * x)$. Furthermore, $x \in X(V)$ if and only if $\beta * x \in X(v_\beta)$. Hence, $x \in \ker(V)$ if and only if $\beta * x \in \ker(v_\beta)$. ∎

Another important class of NTU games are the pure bargaining games. A game $V \in \Gamma$ is a pure bargaining game (PBG) if $V(S) = -\Re^S_+$ for each $S \neq N$. The subset of Γ consisting of PBG's will be denoted Γ^{PB}. If $V \in \Gamma^{PB}$, then $\bigcap_{S \subseteq N} \operatorname{dom} \sigma_{V(S)} = \operatorname{dom} \sigma_{V(N)}$ and $s_{ij}(x, \alpha) > s_{ji}(x, \alpha)$ if and only if $\alpha_i x_i < \alpha_j x_j$. Therefore, we can express the kernel of $V \in \Gamma^{PB}$ as follows: $\ker(V) = \{x \in X(V) |$ for each i, j, $x_j = 0$ or there exists $0 \neq \alpha \in \operatorname{dom} \sigma_{V(N)}$ such that $\alpha_i x_i \geq \alpha_j x_j\}$.

As a result, the kernel of a PBG tends to "shrink" as $\operatorname{dom} \sigma_{V(N)}$ gets smaller. More precisely, if $\operatorname{dom} \sigma_{V(N)} = \Re^n_+$ (as in the case of compactly generated games), then $\ker(V) = X(V)$. However, if $\operatorname{dom} \sigma_{V(N)} = \{t\chi | t \geq 0\}$, then $\ker(V)$ consists of the single point $\{t\chi | t \geq 0\} \cap X(V)$. The kernels of PBG's are related to another well known solution concept. For each $V \in \Gamma^{PB}$ define the Nash Bargaining Solution for V, denoted $\mu(V)$, as follows:

$$\mu(V) = \underset{x \in X(V)}{\arg\max} \prod_{i \in N} x_i.$$

The Nash Bargaining Solution was defined and axiomatically characterized by Nash [20]. The next result relating the kernel and Nash's Bargaining Solution is a consequence of the fact that the NTU kernel of a PBG consists precisely of the Nash Bargaining Solution.

Proposition 5.3. *Suppose* $V \in \Gamma^{BP}$ *and* $\operatorname{dom} \sigma_{V(N)} \setminus \{0\} \subseteq \Re^n_{++}$. *Then* $\mu(V) \in \ker(V)$.

Proof: Let V be as above. It suffices to prove that NTU-$\ker(V) = \{\mu(V)\}$ since the result then follows from Proposition 4.1. To see this, let $x \in$ NTU-$\ker(V)$. Then $x \in X(V)$ and there exists $0 \neq \lambda \in \operatorname{dom} \sigma_{V(N)}$ such that $\lambda * x \in \ker(v_\lambda)$, where $v_\lambda(N) = \sigma_{V(N)}(\lambda)$ and $v_\lambda(S) = 0$ otherwise. If $\sigma_{V(N)}(\lambda) = 0$, then the problem is trivial so suppose that $\sigma_{V(N)}(\lambda) > 0$. In this case, it is easy to verify that the kernel of v_λ is the singleton set $\ker(v_\lambda) = \{(v_\lambda(N)/n)\chi\}$. Alternatively, one could deduce this from the "symmetry" property of the kernel of a TU game or the fact that the kernel and the prekernel coincide for 0-monotonic TU games (see [18]). Hence, we conclude that $\lambda * x \in \ker(v_\lambda)$ if and only if $\Sigma_{i \in N} \lambda_i x_i = v_\lambda(N)$ and $\lambda_i x_i = \lambda_j x_j$ for all $i, j \in N$. Since $\lambda \in \Re^n_{++}$, it follows from standard methods of convex programming [22] (Theorem 27.4) or [25] that these conditions hold if and only if $x = \mu(V)$ and the proof is complete. ∎

The kernel is also related to another well known solution for PBG's. For each $w \in \Re^n_{++}$ and each $V \in \Gamma^{PB}$, define the set of w-egalitarian payoffs for V, denoted E^w, as:

$$E^w(V) = \{x \in X(V) | \text{ for all } i,j \in N, x_i/w_i = x_j/w_j\}.$$

The w-egalitarian solution was defined for PBG's by Kalai [12] (where it is called the "proportional solution") and extended to the general NTU case by Kalai and Samet [13]. The relationship between the kernel and the egalitarian solution is formalized in the next result, whose proof is omitted.

Proposition 5.4. *If* $V \in \Gamma^{PB}$, $w^{-1} \in \Re^n_{++} \cap \operatorname{dom} \sigma_{V(N)}$ *and* $x \in E^w(V)$, *then* $x \in \ker(V)$ *(where* w^{-1} *denotes the vector whose ith coordinate is* $(w^{-1})_i := 1/w_i$*).*

6. References

[1] J.-P. Aubin and I. Ekeland, *Applied Nonlinear Analysis*, John Wiley, New York, 1984.

[2] R. J. Aumann, "An axiomatization of the nontransferable utility value," *Econometrica* **53** (1985) 599–612.

[3] R. J. Aumann and M. Maschler, "The bargaining set for cooperative games," in: *Advances in Game Theory*, M. Dresher, L. S. Shapley, and A. W. Tucker, Eds., *Annals of Mathematics Studies* **52** (1964) 443–476.

[4] N. Asscher, "An ordinal bargaining set for games without side payments," *Mathematics of Operations Research* **1** (1976) 381–389.

[5] L. J. Billera, "Existence of general bargaining sets for cooperative games without side payments," *Bulletin of the American Mathematical Society* **76** (1970) 375–379.

[6] L. J. Billera, "Some theorems on the core of an n-Person Game," *SIAM Journal on Applied Mathematics* **18** (1970) 567–579.

[7] L. J. Billera, "A note on a kernel and the core of a game without side payments," Technical Report #152, Department of Operations Research, Cornell University, 1972.

[8] E. Baudier, "Competitive equilibrium in a game," *Econometrica* **41** (1973) 1049–1068.

[9] M. Davis and M. Maschler, "The kernel of a cooperative game," *Naval Research Logistics Quarterly* **12** (1965) 223–259.

[10] S. Hart, "An axiomatization of Harsanyi's nontransferable utility value," *Econometrica* **53** (1985) 1295–1313.

[11] E. Kalai, "Excess functions for games without side payments," *SIAM Journal on Applied Mathematics* **29** (1975) 60–71.

[12] E. Kalai, "Proportional solutions to bargaining situations: Interpersonal utility comparisons. *Econometrica* **45** (1977) 1623–1630.

[13] E. Kalai and D. Samet, "Monotonic solutions to general cooperative games," *Econometrica* **53** (1985) 307–327.

[14] R. Kern, "The Shapley transfer value without zero weights. *International Journal of Game Theory* **14** (1985) 73–92.

[15] M. Maschler B. Peleg, "A Characterization, existence proof and dimension bounds for the kernel of a Game," *Pacific Journal of Mathematics* **18** (1966) 289–328.

[16] M. Maschler and B. Peleg, "The structure of the kernel of a cooperative Game, *SIAM Journal on Applied Mathematics* **15** (1967) 569–604.

[17] M. Maschler, B. Peleg and, L. S. Shapley, "The kernel and bargaining set for convex games," *International Journal of Game Theory* **1** (1972) 73–93.

[18] M. Maschler, B. Peleg, and L. S. Shapley, "Geometric properties of the kernel, nucleolus and related solution concepts," *Mathematics of Operations Research* **4** (1979) 303–338.

[19] M. Nakayama, "A note on a generalization of the nucleolus to games without sidepayments," *International Journal of Game Theory* **12** (1983) 115–122.

[20] J. Nash, "The bargaining problem," *Econometrica* **28** (1950) 155–162.

[21] B. Peleg, "Bargaining sets of cooperative games without side payments," *Israel Journal of Mathematics* **1** (1963) 197–200.

[22] R. T. Rockafellar, *Convex Analysis*, Princeton University Press, Princeton, New Jersey, 1970.

[23] D. Schmeidler, "The nucleolus of a characteristic function Game," *SIAM Journal on Applied Mathematics* **17** (1969) 1163–1170.

[24] L. S. Shapley, "On balanced sets and cores," *Naval Research Logistics Quarterly* **14** (1967) 453–460.

[25] L. S. Shapley, "Utility comparison and the theory of games," in: *La Decision Paris: Editions du Centre National de la Recherche Scientifique*, 1969.

Essay 5

Relationships Between Bankruptcy Games and Minimum Cost Spanning Tree Games

Theo S.H. Driessen[1]

ABSTRACT. The cost allocation problem of setting charges to the users in a given network is considered with the aid of minimum cost spanning trees in the network. In addition, the bankruptcy problem of dividing the estate of the bankrupt concern among the various creditors according to their claims is treated. The paper aims to establish that, under certain circumstances, the bankruptcy problem can be seen as a minimum cost spanning tree allocation problem. In the game theoretic setting, the paper provides the conditions on the estate and the claims which are sufficient and necessary for the representation of bankruptcy games as minimum cost spanning tree games. Finally, it is shown that a particular cost allocation rule for the minimum cost spanning tree allocation problem agrees with a certain division rule for the bankruptcy problem.

1. Introduction

Consider the situation in which one central source (the common supplier) provides institutions or individuals (the users) with certain goods or services. When once the installation costs of the various connections between two different users or a user and the common supplier are determined, they cooperate in the undertaking of a joint installation of a network of connections in order to supply all users. The least cost of undertaking such a joint venture can be achieved by the construction of a minimum cost spanning tree in the complete graph. Now the common supplier is confronted with the cost allocation problem of setting charges to the users. How to allocate the least total installation cost to the users? The game theoretic approach

---

[1]University of Twente, Department of Applied Mathematics, P.O. Box 217, 7500 AE Enschede, The Netherlands

to the cost allocation problem is based on the notion of a cooperative game in characteristic function form. Let N be the set of users. The function c is defined for any subset S of users such that $c(S)$ represents the least installation cost of a network of connections in order to supply only the members of S, *i.e.*, the cost of a minimum cost spanning tree on S together with the common supplier. This type of a game $\langle N, c\rangle$ is called a minimum cost spanning tree game. As a matter of fact, the least cost of the joint venture equals $c(N)$ and the game theoretic analysis to the cost allocation problem deals with the study of fair divisions of the cost $c(N)$ among the users. The cost figures $c(S)$, $S \subset N$, are involved to conclude whether or not a proposed division of $c(N)$ is fair. With the given cost function c we associate the cost savings function v by means of $v(S) := \sum_{i \in S} c(\{i\}) - c(S)$ for any subset S of users. Here the expression $v(S)$ represents the cost savings that would result from cooperation between the members of S instead of acting alone. Evidently, there are no cost savings for individuals, *i.e.*, $v(\{i\}) = 0$ for any user $i \in N$. In view of the cost savings function, one can say the cost allocation problem is similar to the division problem in terms of savings.

A peculiar example of the division problem is the bankruptcy problem. The data of the bankruptcy situation are given by the estate of the bankrupt concern and the claims of the various creditors. The problem is that the claims are mutually inconsistent in that the estate is insufficient to meet all of the claims. The bankruptcy problem can be formulated as a game theoretic division problem by modelling the bankruptcy situation as a cooperative savings game. Let N be the set of creditors. The savings function u is defined for any subset S of creditors such that $u(S)$ represents the remainder of the estate after each creditor outside S is paid his claim. This type of a game $\langle N, u\rangle$ is called a bankruptcy game. Here the estate equals $u(N)$ and the division problem focuses on fair divisions of the savings $u(N)$ among the creditors.

The aim of the paper is to establish that, under certain circumstances, the bankruptcy problem can be regarded as a minimum cost spanning tree allocation problem. In other words, it is proved that some bankruptcy games can be interpreted as minimum cost spanning tree games. The organization of the paper is as follows. The exact definitions of both the minimum cost spanning tree game and the bankruptcy game are given in Section 2. The main two theorems in Section 3 provide the conditions on the estate and the claims which are sufficient and necessary for the representation of bankruptcy games as minimum cost spanning tree games. In Section 4 we treat the relationship of a specific cost allocation rule (based on minimum cost spanning trees) to a certain division rule for the bankruptcy problem.

2. Minimum cost spanning tree games and bankruptcy games

Concerning the network situation mentioned at the beginning of Section 1, let $N = \{1, 2, \ldots, n\}$ be the set of users and denote the common supplier by 0. It is supposed that the complete graph with node set $N \cup \{0\}$ and undirected edge set $E(N) := \{(i,j) \mid i,\ j \in N \cup \{0\},\ i \neq j\}$ is equipped with a nonnegative weight function $w : E(N) \to \Re_+$. The weight $w((i,j))$ is interpreted as the installation cost of the connection between the two nodes i and j, or equivalently, of the undirected edge (i,j).

Let $\mathcal{P}(N)$ denote the power set of N. The network situation above can be modelled as a cooperative cost game $\langle N, c\rangle$ whose characteristic cost function $c : \mathcal{P}(N) \to \Re$ is given by

$$c(S) := \sum_{e \in E_S} w(e) \qquad \text{for all } S \subset N$$

where E_S represents the edge set of any minimum cost spanning tree on node set $S \cup \{0\}$. Thus, $c(S)$ equals the least total installation cost to connect merely the members of S with the common supplier. The so-called minimum cost spanning tree (*mcst*) game $\langle N, c\rangle$ was introduced in [4] and studied in [1], [2], [7], [8], [9]. Notice that $c(\{i\}) = w((0,i))$ for user $i \in N$. In case the installation cost of the connection between two users is larger than the maximum of the installation costs of their separate connections with the common supplier, then the relevant connection between themselves never occurs in a minimum cost spanning tree on any node set containing both users and the common supplier. Consequently, without loss of generality we may suppose that the weight function w satisfies the condition $w((i,j)) \leq \max[w((0,i)),\ w((0,j))]$ for all $i,j \in N$, $i \neq j$. This implies that $c(\{i,j\}) = w((i,j)) + \min[w((0,i)),\ w((0,j))]$ for all $i,j \in N$, $i \neq j$, where the cost figure $c(\{i,j\})$ arises from a minimum cost spanning tree on node set $\{0,i,j\}$ and containing the edge (i,j).

Concerning the bankruptcy situation mentioned in the middle of Section 1, let $N = \{1, 2, \ldots, n\}$ be the set of creditors and denote the estate of the bankrupt concern and the claims of the n creditors by $E, d_1, d_2, \ldots, d_n$ respectively. Without loss of generality, these n claims may be ordered so that $0 < d_1 \leq d_2 \leq \ldots \leq d_n$ and furthermore, the estate E satisfies $0 \leq E < \sum_{i \in N} d_i$ (otherwise, the bankruptcy problem would not exist).

The bankruptcy situation above can be modelled as a cooperative savings game $\langle N, u\rangle$ whose characteristic savings function $u : \mathcal{P}(N) \to \Re$ is given by

$$u(S) := \max\left[0, E - \textstyle\sum_{i \in N \setminus S} d_i\right] \qquad \text{for all } S \subset N.$$

Thus, $u(S)$ equals either zero or what is left of the estate after each creditor outside S is paid his claim. The so-called bankruptcy game $\langle N, u\rangle$ was introduced in [10] and studied in [3], [5]. Note that $u(N) = E$. For notational convenience we rewrite the characteristic function $u : \mathcal{P}(N) \to \Re$ as follows:

$$u(S) = \max\left[0, \textstyle\sum_{i\in S} d_i - D\right] \qquad \text{for all } S \subset N \tag{2.1}$$

where the positive number $D := \sum_{i\in N} d_i - E$ represents the part of the claims that can not be met by the estate. Generally speaking, the bankruptcy game $\langle N, u\rangle$ fails to be a zero normalized game, *i.e.*, $u(\{i\}) = 0$ for all $i \in N$. Throughout the remainder of the paper, we replace the bankruptcy game $\langle N, u\rangle$ by its zero normalized version $\langle N, v\rangle$ which is defined by

$$v(S) := u(S) - \sum_{i\in S} u(\{i\}) \qquad \text{for all } S \subset N. \tag{2.2}$$

The main goal is to prove that, under certain conditions on the estate and the claims, the zero normalized bankruptcy game of (2.2) can be interpreted as the cost savings game arising from a minimum cost spanning tree game. In other words, given the game $\langle N, v\rangle$ of (2.2), we look for a *mcst* game $\langle N, c\rangle$ satisfying $v(S) = \sum_{i\in S} c(\{i\}) - c(S)$ for all $S \subset N$.

Concerning properties of bankruptcy games and *mcst* games we remark the following. It is well-known that the bankruptcy game $\langle N, u\rangle$ possesses the convexity property, *i.e.*,

$$u(S \cup \{i\}) - u(S) \leq u(T \cup \{i\}) - u(T)$$
$$\text{for all } i \in N \text{ and all } S, T \subset N \text{ such that } S \subset T \subset N \setminus \{i\}.$$

The verification of the convexity condition can easily be done by means of the elegant formula for the characteristic function u of the bankruptcy game. Generally speaking, cost savings games arising from *mcst* games don't possess the convexity property. As a matter of fact, general properties of *mcst* games (or a nice description of the class of *mcst* games) are very difficult to encounter because of the unpredictability of the structure of the *mcst*'s on the various node sets $S \cup \{0\}$, $S \subset N$. The structure of a *mcst* on any node set $S \cup \{0\}$ is not at all related to that of a *mcst* on another node set $T \cup \{0\}$, even if $S \subset T$.

In view of the convexity condition, the representation of the zero normalized bankruptcy game as the cost savings game arising from a *mcst* game is not true in general. The forthcoming Theorems 3.2 and 3.4 provide sufficient and necessary conditions for the relevant representation. We remark already that the proofs of both theorems are indeed exhaustive, due to the unpredictability of the structure of the *mcst*'s and the inapplicable description of the total weight of those *mcst*'s. This is also the reason to study one *mcst* game at a time and to avoid the study of a subclass of *mcst* games and related bankruptcy games.

Example 2.1. Look at the bankruptcy problem with the three claims $d_1 = 100$, $d_2 = 150$, $d_3 = 175$, and the estates $E = 300$ and $E = 200$ respectively.

(i) Suppose $E = 300$. Then the three-person bankruptcy game $\langle N, v\rangle$ of (2.2) is given by $v(\{i\}) = 0$ for all $i \in N$, $v(\{1,2\}) = v(\{1,3\}) = 100$, $v(\{2,3\}) = 125$ and $v(N) = 225$. It appears that the game $\langle N, v\rangle$ can be associated with the network situation whose weight function w on the edge set is given by

$$\begin{array}{lll} w((0,1)) = 100, & w((0,2)) = 50, & w((0,3)) = 125, \\ w((1,2)) = 0, & w((1,3)) = 25, & w((2,3)) = 0\ . \end{array}$$

The *mcst* game $\langle N, c\rangle$ corresponding to this network is as follows:

$$\begin{array}{lll} c(\{1\}) = 100, & c(\{2\}) = 50, & c(\{3\}) = 125, \\ c(\{1,2\}) = 50, & c(\{1,3\}) = 125, & c(\{2,3\}) = 50, \end{array}$$

and $c(N) = 50$. Now it is straightforward to verify that the associated cost savings game is precisely the game $\langle N, v\rangle$. Notice that there exist other *mcst* games whose associated cost savings game coincides with the bankruptcy game $\langle N, v\rangle$. For instance, those *mcst* games arising from the network with a slightly adapted weight function in the sense that the original weight of each edge is increased by the same positive amount. The latter reasoning is true in general.

(ii) Suppose $E = 200$. Then the three-person bankruptcy game $\langle N, v\rangle$ of (2.2) is given by $v(\{i\}) = 0$ for all $i \in N$, $v(\{1,2\}) = 25$, $v(\{1,3\}) = 50$, $v(\{2,3\}) = 100$ and $v(N) = 200$. Now we assert that there exists no *mcst* game $\langle N, c\rangle$ whose associated cost savings game equals the game $\langle N, v\rangle$. If such a *mcst* game $\langle N, c\rangle$ would exist, then the game should satisfy the following four equalities:

$$\begin{aligned} c(\{1,2\}) &= c(\{1\}) + c(\{2\}) - 25 \\ c(\{1,3\}) &= c(\{1\}) + c(\{3\}) - 50 \\ c(\{2,3\}) &= c(\{2\}) + c(\{3\}) - 100 \\ c(\{1,2,3\}) &= c(\{1\}) + c(\{2\}) + c(\{3\}) - 200\ . \end{aligned}$$

We establish that the fourth equality cannot be met by a *mcst* game $\langle N, c\rangle$ satisfying the first three equalities. Let us consider the weight function w in the network situation involving the *mcst* game $\langle N, c\rangle$. Write $N = \{i, j, k\}$ where $w((0,i)) \le w((0,j)) \le w((0,k))$. Then we obtain

$$\begin{aligned} w((i,j)) &= c(\{i,j\}) - \min[w((0,i)), w((0,j))] \\ &= c(\{i\}) + c(\{j\}) - v(\{i,j\}) - w((0,i)) \\ &= w((0,j)) - v(\{i,j\}) \\ &\le w((0,j)) \end{aligned}$$

and similarly,

$$\begin{aligned} w((i,k)) &= w((0,k)) - v(\{i,k\}) \leq w((0,k)), \\ w((j,k)) &= w((0,k)) - v(\{j,k\}) \leq w((0,k)). \end{aligned}$$

With this description of the weight function w at hand and checking the various possibilities for a minimum cost spanning tree on node set $\{0,1,2,3\}$, we arrive at the conclusion that the total cost $c(N)$ equals

$$\begin{aligned} \text{either} \quad & w((0,i)) + w((i,j)) + w((i,k)), \\ & w((0,i)) + w((i,j)) + w((j,k)), \\ \text{or} \quad & w((0,i)) + w((i,k)) + w((k,j)). \end{aligned}$$

Therefore,

$$c(N) \geq w((0,1)) + w((0,2)) + w((0,3)) - 150 = \sum_{\ell \in N} c(\{\ell\}) - 150.$$

This, however, is in contradiction with the fourth equality mentioned at the beginning of (ii).

3. Representation of bankruptcy games as minimum cost spanning tree games

We are interested in the possible representation of the zero normalized bankruptcy game as the cost savings game arising from a *mcst* game. In point of fact, we attempt to present the conditions (on the estate and the claims) which are sufficient and necessary for the relevant relationship between zero normalized bankruptcy games and *mcst* games. Firstly, we treat an appropriate reformulation of the characteristic function of the zero normalized bankruptcy game. Given the claims d_i, $i \in N$, of the creditors, put $d := (d_1, d_2, \ldots, d_n) \in \Re^n$. For any vector $x \in \Re^n$ and any coalition $S \subset N$, we write $x(S)$ instead of $\sum_{i \in S} x_i$ throughout this section. The number of members of S is denoted by $|S|$.

Lemma 3.1. *Let $\langle N, v \rangle$ be the zero normalized bankruptcy game of (2.2).*

(i) $v(S) = \sum_{i \in S} \min[d_i, D] - \min[d(S), D]$ *for all* $S \subset N$.

(ii)

$$v(S) = \begin{cases} 0 & \text{if } S \subset N \text{ with } d(S) \leq D \\ \sum_{i \in S} \min[d_i, D] - D & \text{if } S \subset N \text{ with } d(S) \geq D \;. \end{cases}$$

(iii) If $D \leq d_1$, then $v(S) = (|S|-1)D$ for all $S \subset N$, $S \neq \emptyset$.

Proof: For any $S \subset N$, it follows immediately from (2.1) – (2.2) that

$$\begin{aligned} v(S) &= \max[0, d(S) - D] - \sum_{i \in S} \max[0, d_i - D] \\ &= d(S) + \max[-d(S), -D] - \sum_{i \in S} [d_i + \max[-d_i, -D]] \\ &= -\min[d(S), D] + \sum_{i \in S} \min[d_i, D] \ . \end{aligned}$$

This proves part (i). Part (ii) is a direct consequence of part (i) by noting that $d(S) \leq D$ implies $d_i \leq d(S) \leq D$ for all $i \in S$. Part (iii) is obtained from part (ii) by noting that $d_1 \geq D$ implies $d_i \geq d_1 \geq D$ for all $i \in N$ as well as $d(S) \geq D$ for all $S \subset N$, $S \neq \emptyset$. ∎

Theorem 3.2. *The zero normalized bankruptcy game $\langle N, v\rangle$ of (2.2) can be represented as the cost savings game arising from a* mcst *game with user set N if one of the following two conditions is satisfied:*

(i) $0 < D \leq d_3$

(ii) There exists a unique $k \in \{3, 4, \ldots, n-1\}$ such that

$$d_k < D \leq d_{k+1} \qquad \textit{and} \qquad \sum_{i=1}^{k} d_i \leq D. \tag{3.1}$$

Proof: Consider the complete graph with node set $N \cup \{0\}$ and undirected edge set $E(N)$. Define the weight function $w : E(N) \to \Re_+$ by

$$\begin{aligned} w((0,i)) &:= 2D \qquad \text{for all } i \in N, \\ w((i,j)) &:= \begin{cases} D & \text{if} \quad D \leq d_i \leq d_j \\ 2D - d_i & \text{if} \quad d_i \leq D \leq d_j \\ 2D - \max[0, d_i + d_j - D] & \text{if} \quad d_i \leq d_j \leq D \ . \end{cases} \end{aligned}$$

Note that $D \leq w((i,j)) \leq 2D$ for all $i, j \in N$, $i \neq j$. Let $\langle N, c\rangle$ be the *mcst* game corresponding to the network situation involved. Evidently, $c(\{i\}) = w((0,i)) = 2D$ for all $i \in N$ and further, $c(\{i,j\}) = w((0,i)) + w((i,j)) = 2D + w((i,j))$ for all $i, j \in N, i \neq j$. We show that the equality $\sum_{i \in S} c(\{i\}) - c(S) = v(S)$ holds for all $S \subset N$, on the understanding that either $0 < D \leq d_3$ or condition (3.1) is satisfied. The relevant equality holds trivially for one-person coalitions S because the game $\langle N, v\rangle$ is zero normalized. Concerning two-person coalitions $S = \{i, j\}$, $i \neq j$, we have

$$c(\{i\}) + c(\{j\}) - c(\{i,j\}) = 2D - w((i,j)) = v(\{i,j\})$$

where the last equality is a direct consequence of Lemma 3.1(i) and the definition of the weight $w((i,j))$. In order to prove that the relevant equality also holds for multi-person coalitions, let $S \subset N$ satisfy $s := |S| \geq 3$. Put $i_0 := 0$ and write $S = \{i_1, i_2, i_3, \ldots, i_s\}$ where $0 := d_{i_0} \leq d_{i_1} \leq d_{i_2} \leq \ldots \leq d_{i_s}$. Now we distinguish two cases.

(i) Suppose $0 < D \leq d_3$. Note that $d_3 \leq d_{i_3}$ because $s \geq 3$. Define the critical number $r = r(S)$ as the smallest number $r \in \{1,2,3\}$ satisfying $d_{i_{r-1}} < D \leq d_{i_r}$. The critical number r is well-defined since $0 < D \leq d_{i_3}$. Due to the construction of the weight function w, we conclude that a minimum cost spanning tree on node set $S \cup \{0\}$ is given by the following spanning tree with edges $(0, i_r)$ and (i_r, i_j) for all $1 \leq j \leq s$, $j \neq r$. The weights of these edges are described by

$$\begin{aligned} w((0,i_r)) &= 2D \\ w((i_r,i_j)) &= \begin{cases} 2D - d_{i_j} & \text{if} \quad 1 \leq j \leq r-1 \\ D & \text{if} \quad r+1 \leq j \leq s. \end{cases} \end{aligned}$$

The reader is invited to sketch this minimum cost spanning tree. From this we deduce

$$c(S) = \sum_{j=1}^{r-1} (2D - d_{i_j}) + 2D + (s-r)D = (s+r)D - \sum_{j=1}^{r-1} d_{i_j} \,.$$

Hence,

$$\sum_{i \in S} c(\{i\}) - c(S) = 2sD - c(S) = (s-r)D + \sum_{j=1}^{r-1} d_{i_j} = v(S)$$

where the last equality is a direct consequence of Lemma 3.1(i) with respect to the coalition S involved. This completes the first case.

(ii) Suppose (3.1) holds. If $d(S) \leq D$, then $d_i \leq d_i + d_j \leq d(S) \leq D$ and so, $w((i,j)) = 2D$ for all $i,j \in S$, $i \neq j$. Thus, if $d(S) \leq D$, then any spanning tree on node set $S \cup \{0\}$ is a minimum cost spanning tree on $S \cup \{0\}$ with associated cost $c(S) = 2sD$ and therefore, the cost savings are given by $\sum_{i \in S} c(\{i\}) - c(S) = 2sD - 2sD = 0 = v(S)$. Here the last equality follows from Lemma 3.1(ii). It remains to consider the subcase $d(S) > D$. Together with $\sum_{i=1}^{k} d_i \leq D$, this yields that there exists $j \in S$ satisfying $j \geq k+1$. Define the critical number $r = r(S)$ as the smallest number $r \in \{1, 2, \ldots, s\}$ satisfying $d_{i_r} \geq d_{k+1}$. Since $d_k < D \leq d_{k+1}$, we also have $d_{i_{r-1}} < D \leq d_{i_r}$. From now on, the proof of part (i) is once again applicable to complete this subcase $d(S) > D$. ∎

According to the theorem above, the representation of a zero normalized n-person bankruptcy game as the cost savings game arising from a n-person *mcst* game is realizable whenever the estate satisfies the constraints of exactly one of in total $n-2$ closed intervals in $\Re$ whose bounds are given in terms of the claims. These $n-2$ closed intervals are described by

$$[d(N\setminus\{3\}), d(N)], \quad [d(N\setminus\{4\}), d(N\setminus\{1,2,3\})],$$
$$[d(N\setminus\{5\}), d(N\setminus\{1,2,3,4\})], \ldots, [d(N\setminus\{n-1\}), d_{n-1}+d_n],$$
$$\text{and } [d(N\setminus\{n\}),\ d_n]\ .$$

Concerning zero normalized three- or four-person bankruptcy games, the relevant representation is realizable whenever the estate is large enough (*i.e.*$E \geq d_1+d_2$ for $n=3$ and $E \geq d_1+d_2+d_4$ for $n=4$) or sufficiently large (*i.e.*$d_1+d_2+d_3 \leq E \leq d_4$ for $n=4$). Remark that the sufficient condition $E \geq d_1+d_2$ for $n=3$ was fulfilled in Example 2.1(i), but not in Example 2.1(ii). The next example illustrates that the relevant representation is not necessarily realizable whenever $n=4$ and $d_4 < E < d_1+d_2+d_4$.

Example 3.3. Look at the bankruptcy problem with the estate $E=190$ and the four claims $d_1=40$, $d_2=60$, $d_3=80$, $d_4=100$. The corresponding zero normalized four-person bankruptcy game $\langle N, v\rangle$ of (2.2) is given by

$$\begin{array}{lll} v(\{1,2\})=10, & v(\{2,3\})=50, & v(\{1,2,3\})=90, \\ v(\{1,3\})=30, & v(\{2,4\})=60, & v(\{1,2,4\})=100, \\ v(\{1,4\})=40, & v(\{3,4\})=80, & v(\{1,3,4\})=120, \\ & & v(\{2,3,4\})=140, \end{array}$$

and $v(N)=180$. Notice that $D=90$ and so, $d_3 < D < d_4$. The condition (3.1), however, fails to hold since $d_1+d_2+d_3 > D$. Now we assert that there exists no *mcst* game $\langle N, c\rangle$ whose associated cost savings game equals the game $\langle N, v\rangle$. The main bottleneck is caused by the coalition $\{1,2,3\}$. If such a *mcst* game $\langle N, c\rangle$ would exist, then the game should satisfy, among others, the equalities

$$\begin{aligned} c(\{1,2\}) &= c(\{1\})+c(\{2\})-10, \\ c(\{1,3\}) &= c(\{1\})+c(\{3\})-30, \\ c(\{2,3\}) &= c(\{2\})+c(\{3\})-50, \\ c(\{1,2,3\}) &= c(\{1\})+c(\{2\})+c(\{3\})-90\ . \end{aligned}$$

Without going into technical details, it appears that the fourth equality cannot be met by a *mcst* game satisfying the first three equalities (cf. the approach mentioned in Example 2.1(ii)). Notice that, if the fourth claim is replaced by $d_4=200$, then condition (3.1) holds and consequently, the adapted bankruptcy game can be represented by means of a *mcst* game.

Theorem 3.4. *The zero normalized bankruptcy game $\langle N, v\rangle$ of (2.2) cannot be represented as the cost savings game arising from a* mcst *game with user set N if one of the following two conditions is satisfied:*

(i) $D > d_n$

(ii) There exists a unique $k \in \{3, 4, \ldots, n-1\}$ *such that*

$$d_k < D \leq d_{k+1} \qquad \text{and} \qquad \sum_{i=1}^{k} d_i > D. \tag{3.2}$$

Proof: Assume, on the contrary, that such a *mcst* game $\langle N, c\rangle$ would exist, on the understanding that either $D > d_n$ or condition (3.2) is satisfied. Let $w : E(N) \to \Re_+$ denote the weight function in the network situation involving the *mcst* game $\langle N, c\rangle$. Then the cost savings for any two users $i, j \in N$ are given by $v(\{i,j\}) = c(\{i\}) + c(\{j\}) - c(\{i,j\})$ and hence, the cost $c(\{i,j\}) = w((0,i)) + w((0,j)) - v(\{i,j\})$. From this and Lemma 3.1(i), we deduce that for all $i, j \in N$, $i \neq j$,

$$c(\{i,j\}) = w((0,i)) + w((0,j)) - \min[d_i, D] - \min[d_j, D] + \min[d_i + d_j, D]$$

and therefore,

$$w((i,j)) = c(\{i,j\}) - \min[w((0,i)), w((0,j))] =$$

$$\max[w((0,i)), w((0,j))] + \min[d_i + d_j, D] - \min[d_i, D] - \min[d_j, D] \,.$$

To be exact, for any $i, j \in N$, $i \neq j$, satisfying $d_i \leq d_j \leq D$, we obtain

$$w((i,j)) = \max[w((0,i)), w((0,j))] - \max[0, d_i + d_j - D]. \tag{3.3}$$

Put $d_{n+1} := +\infty$. From now on, suppose that there exists a unique $k \in \{3, 4, \ldots, n\}$ such that $d_k < D \leq d_{k+1}$ and $\sum_{i=1}^{k} d_i > D$. Fix the coalition $S_k := \{1, 2, \ldots, k\}$. Then $d_i \leq d_k < D$ for all $i \in S_k$ as well as $d(S_k) > D$. From this and Lemma 3.1(i), we obtain that the cost $c(S_k)$ must satisfy

$$c(S_k) = \sum_{i \in S_k} c(\{i\}) - v(S_k) = \sum_{i \in S_k} w((0,i)) - d(S_k) + D. \tag{3.4}$$

If $d_i + d_j \leq D$ for all $i, j \in S_k$, $i \neq j$, then $w((i,j)) = \max[w((0,i)), w((0,j))]$ for all $i, j \in S_k$, $i \neq j$, and consequently, the cost of a minimum cost spanning tree on $S_k \cup \{0\}$ would be given by $c(S_k) = \sum_{i \in S_k} w((0,i))$. The latter equality, however, is in contradiction with (3.4) because $d(S_k) > D$. Therefore, we may assume that there exists at least one pair $i_1, i_2 \in S_k$, $i_1 \neq i_2$, satisfying $d_{i_1} + d_{i_2} > D$. Since $|S_k| = k \geq 3$, we can choose an arbitrary $i_3 \in S_k \setminus \{i_1, i_2\}$. Put $T := \{i_1, i_2, i_3\}$. By using again Lemma 3.1(i), we obtain on the one side

$$c(T) = \sum_{i \in T} c(\{i\}) - v(T) = \sum_{i \in T} w((0,i)) - d(T) + D. \tag{3.5}$$

On the other, given the description (3.3) of the weight function w and checking the various possibilities for a *mcst* on node set $T \cup \{0\}$, we arrive at the conclusion that the total cost of the three edges in a *mcst* on $T \cup \{0\}$ is at least the amount of $\sum_{i \in T} w((0,i))$ minus two times a positive term of the form $d_i + d_j - D$ where $i, j \in T$, $i \neq j$. This reasoning, together with $D > d_i$ for all $i \in T$, yields that the cost $c(T)$ satisfies the strict inequality $c(T) > \sum_{i \in T} w((0,i)) - d(T) + D$. Now the contradiction with (3.5) completes the proof. ∎

In view of the two Theorems 3.2 and 3.4, the conclusion is that the zero normalized bankruptcy game can be represented as the cost savings game arising from a *mcst* game if and only if either $0 < D \leq d_3$ or condition (3.1) holds. These sufficient and necessary conditions for the relevant representation are already interpreted after the proof of Theorem 3.2.

4. Comparison between minimum cost spanning tree allocation rules and division rules for bankruptcy problems

The purpose is to establish that a particular solution for the minimum cost spanning tree allocation problem agrees with a certain solution for the bankruptcy problem. Consider an arbitrary *mcst* game $\langle N, c \rangle$. Let Γ_N be any minimum cost spanning tree on node set $N \cup \{0\}$. The idea of the solution for the *mcst* allocation problem is the following. In order to allocate the total weight of edges in the tree Γ_N, each edge in Γ_N is associated with exactly one of its end points (the one whose indirect connection in the tree with the common supplier uses the other end point) and subsequently, the installation cost of each edge in Γ_N is charged to the user located at the designated end point of the edge. Formally, the *mcst* allocation rule is defined as follows. Take a *mcst* Γ_N on node set $N \cup \{0\}$. For any user $i \in N$, determine the immediate predecessor $p_i(\Gamma_N)$ of user i on the unique path from user i to the common supplier 0 in the tree Γ_N and allocate the weight $w((p_i(\Gamma_N), i))$ to user i. Due to its construction, the *mcst* allocation rule charges the total amount of $c(N)$ to all users.

The next theorem states that, under certain circumstances, the *mcst* allocation rule can be associated with a specific division rule for the bankruptcy problem. The rule in question divides the estate of the bankrupt concern among the creditors according to their claims except for one creditor, whose nonnegative payoff equals his claim lowered by the amount D.

Theorem 4.1. *Suppose that the bankruptcy problem with estate E and claims d_1, $d_2, \ldots, d_n$ ($d_0 := 0$) can be represented as a* mcst *allocation* *problem. Then the* mcst *allocation rule agrees with the following division rule for the bankruptcy problem:*

$$(d_1, d_2, \ldots, d_k, d_{k+1} - D, d_{k+2}, \ldots, d_n)$$

where $k \in \{0, 1, 2, \ldots, n-1\}$ *is determined by* $d_k < D \leq d_{k+1}$.

Proof: Let the zero normalized bankruptcy game $\langle N, v\rangle$ of (2.2) be represented as the cost savings game arising from the *mcst* game $\langle N, c\rangle$ indicated in the proof of Theorem 3.2. Since either $0 < D \leq d_3$ or condition (3.1) holds, there exists a unique $k \in \{0, 1, 2, \ldots, n-1\}$ such that $d_k < D \leq d_{k+1}$. Look once again at the proof of Theorem 3.2 with respect to the coalition $S = N$ and $i_j = j$ for all $j \in N \cup \{0\}$. In both cases (i) and (ii), the critical number $r = r(N)$ is determined by $r = k + 1$. Further, a minimum cost spanning tree on node set $N \cup \{0\}$ is given by the following spanning tree with edges $(0, k+1)$ and $(k+1, i)$ for all $1 \leq i \leq n$, $i \neq k+1$. The weights of these edges are described by

$$\begin{aligned} w((0, k+1)) &= 2D \\ w((k+1, i)) &= \begin{cases} 2D - d_i & \text{if} \quad 1 \leq i \leq k \\ D & \text{if} \quad k+2 \leq i \leq n\,. \end{cases} \end{aligned}$$

The reader is invited to sketch this minimum cost spanning tree. The corresponding *mcst* allocation $y \in \Re^n$ is given by

$$y_i = \begin{cases} 2D - d_i & \text{if} \quad 1 \leq i \leq k \\ 2D & \text{if} \quad i = k+1 \\ D & \text{if} \quad k+2 \leq i \leq n\,. \end{cases}$$

Recall that the bankruptcy game $\langle N, u\rangle$ of (2.1) can be described by means of

$$u(S) = v(S) + \sum_{i \in S} u(\{i\}) = \sum_{i \in S} c(\{i\}) + \sum_{i \in S} u(\{i\}) - c(S)$$

for all $S \subset N$. Thus, the *mcst* allocation y in the *mcst* game $\langle N, c\rangle$ is associated with the division $x \in \Re^n$ in the bankruptcy game $\langle N, u\rangle$ by means of

$$x_i = c(\{i\}) + u(\{i\}) - y_i \qquad \text{for all } i \in N.$$

Straightforward calculations yield that for all $i \in N$

$$\begin{aligned} x_i &= c(\{i\}) + u(\{i\}) - y_i = w((0, i)) + \max[0, d_i - D] - y_i \\ &= 2D + \max[0, d_i - D] - y_i \qquad \text{so,} \end{aligned}$$

$$x_i = \begin{cases} d_i & \text{if} \quad i \neq k+1 \\ d_i - D & \text{if} \quad i = k+1\,. \end{cases}$$

∎

Concerning Theorem 4.1, the proposed division rule for the bankruptcy problem strongly depends upon the choice of the minimum cost spanning tree on node set $N \cup \{0\}$ and merely disadvantages the unique creditor (node $k+1$) who is connected with the common supplier in the given *mcst*. Now it turns out that each other creditor can also be disadvantaged according to the *mcst* allocation rule by adapting the given *mcst*. The least cost is not at all affected whenever the edge $(0, k+1)$ is replaced by an arbitrary edge $(0, i)$, $i \in N$, because the installation cost of each such edge equals $2D$.

In case the edge $(0, k+1)$ in the given *mcst* is replaced by an edge $(0, \widehat{i})$, where $k+2 \leq \widehat{i} \leq n$, then the corresponding *mcst* allocation $y \in \Re^n$ and the associated division $x \in \Re^n$ for the bankruptcy problem are given by

$$(y_i, x_i) = \begin{cases} (2D - d_i, d_i) & \text{if} \quad 1 \leq i \leq k \\ (2D, d_i - D) & \text{if} \quad i = \widehat{i} \\ (D, d_i) & \text{if} \quad k+1 \leq i \leq n, \ i \neq \widehat{i} \ . \end{cases}$$

In case the edge $(0, k+1)$ in the given *mcst* is replaced by an edge $(0, \widehat{i})$, where $1 \leq \widehat{i} \leq k$, then the corresponding *mcst* allocation $y \in \Re^n$ and the associated division $x \in \Re^n$ for the bankruptcy problem are given by

$$(y_i, x_i) = \begin{cases} (2D, 0) & \text{if} \quad i = \widehat{i} \\ (2D - d_i, d_i) & \text{if} \quad 1 \leq i \leq k, \ i \neq \widehat{i} \\ (2D - d_{\widehat{i}}, d_i - D + d_{\widehat{i}}) & \text{if} \quad i = k+1 \\ (D, d_i) & \text{if} \quad k+2 \leq i \leq n \ . \end{cases}$$

In the former case the proposed division rule is disadvantageous for only one creditor, whereas in the second case two creditors receive less than their claims. For a detailed treatment of other division rules for the bankruptcy problem, we refer to Chapter VI in [6].

We conclude the paper with a relationship between the proposed division rules for the bankruptcy problem and the core of the bankruptcy game $\langle N, u \rangle$. As usual, the core $C(u)$ is defined by

$$C(u) := \{x \in \Re^n \mid x(N) = u(N) \text{ and } x(S) \geq u(S) \text{ for all } S \subset N\} \ .$$

We claim that all proposed division rules belong to the core. For example, in the latter case where $1 \leq \widehat{i} \leq k$ the proposed division $x \in \Re^n$ satisfies for all $S \subset N$

$$x(S) = \begin{cases} d(S) - D & \text{if} \quad \widehat{i} \in S, \ k+1 \in S \\ d(S \setminus \{\widehat{i}\}) > d(S) - D & \text{if} \quad \widehat{i} \in S, \ k+1 \notin S \\ d(S) + d_{\widehat{i}} - D > d(S) - D & \text{if} \quad \widehat{i} \notin S, \ k+1 \in S \\ d(S) \geq d(S) - D & \text{if} \quad \widehat{i} \notin S, \ k+1 \notin S \end{cases}$$

and so, $x(S) \geq \max[0, d(S) - D] = u(S)$ for all $S \subset N$. Hence, $x \in C(u)$ in case $1 \leq \hat{i} \leq k$. In case $k+1 \leq \hat{i} \leq n$, then it is evident that $x(S) \geq d(S) - D$ for all $S \subset N$ and thus, $x \in C(u)$.

5. References

[1] H.F.M. Aarts and T.S.H. Driessen, "A survey on minimum cost spanning tree games," Working Paper No. 1003, Department of Applied Mathematics, University of Twente, Enschede, The Netherlands, 1991.

[2] H.F.M. Aarts and T.S.H. Driessen, "On the core-structure of minimum cost spanning tree games," Working Paper No. 1085, Department of Applied Mathematics, University of Twente, Enschede, The Netherlands, 1992.

[3] R. J. Aumann and M. Maschler, "Game theoretic analysis of a bankruptcy problem from the Talmud," *Journal Economic Theory* **36** (1985) 195–213.

[4] C. G. Bird, "On cost allocation for a spanning tree: a game theoretic approach," *Networks* **6** (1976) 335–350.

[5] I. J. Curiel, M. Maschler and S. H. Tijs, "Bankruptcy games," *Zeitschrift für Operations Research A* **31** (1987) 143–159.

[6] T.S.H. Driessen, *Cooperative games, solutions and applications*, Kluwer Academic Publishers, Dordrecht, The Netherlands, 1988.

[7] D. Granot and G. Huberman, "Minimum cost spanning tree games," *Mathematical Programming* **21** (1981) 1–18.

[8] D. Granot and G. Huberman, "The relationship between convex games and minimum cost spanning tree games: a case for permutationally convex games," *SIAM Journal of Algebraic and Discrete Methods* **3** (1982) 288–292.

[9] D. Granot and G. Huberman, "On the core and nucleolus of minimum cost spanning tree games," *Mathematical Programming* **29** (1984) 323–347.

[10] B. O'Neill, "A problem of rights arbitration from the Talmud," *Mathematical Social Sciences* **2** (1982) 345–371.

Essay 6

Non-Zero Sum Repeated Games and Information Transmission

Françoise Forges[1]

1. Introduction

The pioneering work of Aumann, Maschler and Stearns (1968) on infinitely repeated games with incomplete information has obviously given rise to extensions of the Folk theorem. In this paper, we show that their analysis has also provided fundamental hints for the characterization of the solutions of (one-shot) games of information transmission.

The typical results of the *supergames*[2] literature consist of *feasibility theorems*[3] which show in particular that cooperative outcomes may be achieved as equilibrium solutions in this model. In repeated games with *incomplete information*, the dynamic structure enables the players to communicate, so that the generalization of the feasibility theorems is not obvious. The definition of cooperative outcomes is already problematic: Should incentive constraints be taken into account? How should these constraints be formulated in such a rich environment? These difficulties were well identified by Aumann and Maschler as early as in the sixties. They initiated the study of infinitely repeated games with incomplete information by exploring the two-person *zero-sum* games with lack of information on one side (Aumann and Maschler (1966), (1967), (1968)). Zero-sum games are indeed appropriate to study strategic information transmission independently of cooperation effects. They are also essential to the definition of the individually rational levels and to the construction of punishment strategies in non-zero games[4].

---

[1]Chercheur qualifié au F.N.R.S, CORE, Belgique.

[2]Repeated games with complete information (see e.g., the entry "supergames", written by Mertens in the New Palgrave (1987)).

[3]This terminology is borrowed from Myerson (1991, p. 331).

[4]In the underlying framework, the existence of the maxmin, the minmax and optimal strategies is even not guaranteed.

After having fully solved[5] the case of zero-sum repeated games with lack of information on one side, Aumann and Maschler turned to non-zero-sum games of the same class. Together with Stearns (1968), they realized a substantial breakthrough toward a generalization of the Folk theorem in the context of incomplete information. This research was completed by Hart (1985) and Aumann and Hart (1986), who obtained the characterization of the set of all Nash equilibrium payoffs. This result will be presented in details below. Sorin (1983) used the equilibrium scenarios pointed out by Aumann, Maschler and Stearns (1968) to establish the existence of Nash equilibrium in the case of two states of nature. The work of Aumann and Maschler was pursued in many other directions, for instance, zero-sum games with lack of information on *both sides* (see Mertens (1987), chapters 5 and 6 of the forthcoming Handbook of Game Theory and the forthcoming book of Mertens, Sorin and Zamir).

In zero-sum games, an unambiguous solution concept is available, the value, which is unique when it exists. In non-zero-sum games, one can use the Nash equilibrium, as suggested above, but also the *correlated equilibrium* (Aumann (1974), (1987)) or the *communication equilibrium* (e.g. Forges (1986), Myerson (1991), p.261). As already observed, in a repeated game with incomplete information, the players have opportunities to exchange information even if explicit communication devices are not allowed. For instance, the moves can be used as signals, if they are observed after every stage. One may thus hope that correlated equilibria and communication equilibria be payoff-equivalent in repeated games with incomplete information, which would illustrate the "power" of the correlated equilibrium solution concept. Equivalence results of this sort have been obtained in the model introduced by Aumann, Maschler and Stearns (1968), namely infinitely repeated games with lack of information on one side (Forges (1985), (1988)).

In the present paper, we focus on a particular case of this model, where the only role of the informed player is to transmit information to the other one. This framework is appropriate to study signalling strategies of the informed player "under laboratory conditions". A basic idea underlying the Folk theorem is that a sequence of moves can be sustained at an equilibrium if each player has a strategy to punish his opponent when the latter does not follow the prescribed scenario. If the actions of the informed player do not influence directly the payoffs (i.e., influence them only through signalling effects), such punishments do not have to be exhibited. The model can be interpreted as a one-shot decision problem (for the uninformed player) pre-

[5]Aumann and Maschler (1968) obtained an explicit formula for the value and described optimal strategies in the general model of repeated games with lack of information on one side, with arbitrary signalling matrices (to describe the information of the players after every stage).

ceded by a conversation between the two players, namely as a generalized *sender-receiver game*[6] (see Crawford and Sobel (1982) and Myerson (1991), p.283). The game is described formally in Section 2.

The results of Hart (1985) and of Aumann and Hart (1986) enable us to characterize the set of all Nash equilibrium payoffs of the game. Surprising phenomena arise; for instance, new equilibrium payoffs can be achieved if additional stages of conversation are allowed. This feature does not depend on the alphabet of messages available to the sender; a sequence of successive signals (from the informed player) and responses (from the uninformed player) may be needed in order to reach a given equilibrium payoff; in such an equilibrium scenario, at each step, the conversation is pursued as a function of the preceding stages. The fact that an additional signalling step could enlarge the set of solutions was already illustrated by Aumann, Maschler and Stearns (1968) in the general framework of repeated games with lack of information on one side, where the payoffs depend directly on *both* players'decisions. In the simple model considered here, one can even exhibit equilibrium payoffs which can only be achieved if *no deadline* is imposed to the conversation (Forges (1984)). These results over Nash equilibria are stated in details in section 3.

Most of these Nash equilibria (which are thus obtained as solutions of an infinitely repeated game) can be immediately interpreted as *cheap talk* equilibria (Farrell (1988)) of the associated sender-receiver game. As suggested above, in this game, the uninformed player makes a single decision (which determines the payoffs) but *both* players may talk in a costless way. This obviously offers signalling opportunities to the informed player. In a less standard way, the *jointly controlled lotteries* introduced by Aumann, Maschler and Stearns (1968) can be viewed as a sophisticated use of cheap talk by the players. The different cheap-talk equilibrium patterns (including the conversation of "unbounded length") are illustrated in Section 3.5.

In Section 4, correlated equilibria are characterized in a very tractable way, by means of simple communication devices, the *noisy channels*. This is the main step to prove that correlated equilibria are payoff-equivalent to communication equilibria in the model under consideration (Forges (1985)). A consequence of this result is that unlike the Nash equilibrium payoffs, the correlated equilibrium payoffs can always be achieved with one stage of information transmission from the informed player to the uninformed one. In this paper, we try to follow Aumann and Maschler's example in conveying the idea of general procedures by the analysis of examples. To this aim, we insist on the geometric properties of the set of correlated equilibrium payoffs, which is a convex polyhedron. We show in particular

[6] In a generalized sender-receiver gamer, both players can talk before the receiver makes his decision (see below).

that if there are only two states of nature, one can restrict attention to noisy channels with at most four signals, whatever the number of actions of the uninformed player. This enables us to describe the set of *all* correlated equilibrium payoffs in several particular games (in section 4.3).

2. The Model

Let $0 \leq p \leq 1$; $\Gamma(p)$ is defined as the following two-person infinitely repeated game:

Stage 0: One of two possible states of nature Γ_1, Γ_2 is chosen according to the probability distribution $(p, 1-p)$; Γ_k is described by payoff matrices $A^k = [A^k(j)]_{j \in J}$, $B^k = [B^k(j)]_{j \in J}$ for player 1 and player 2 respectively, where J is a (finite) set of actions for player 2 ($|J| \geq 2$). Only player 1 is informed of the "true state of nature" (which will sometimes be referred to as player 1's type).

Stage t ($t = 1, 2, \ldots$): Player 1 sends a message in $\{H, T\}$ ('heads', 'tails') and simultaneously, player 2 makes a decision j_t in J. The stage payoffs are $A^k(j_t), B^k(j_t)$; player 1's message is thus costless.

Given a sequence of moves $(j_t)_{t \geq 1}$, the payoffs in the n-stage game are defined by

$$x_n^k = \frac{1}{n} \sum_{t=1}^{n} A^k(j_t) \quad k = 1, 2$$

$$y_n^k = \frac{1}{n} \sum_{t=1}^{n} B^k(j_t) \quad k = 1, 2$$

for player 1 and player 2 respectively.

A pair of strategies (σ, τ) is a Nash equilibrium of $\Gamma(p)$ if

$$\liminf_{n \to \infty} E_{p,\sigma,\tau}(x_n^k) \geq \limsup_{n \to \infty} E_{p,\sigma',\tau}(x_n^k)$$

for every σ', and similarly for player 2. The equilibrium payoffs are well defined, since they converge. These payoffs are denoted as $(a, \beta) \in \Re^2 \times \Re$: $a = (a^1, a^2)$ is player 1's expected vector payoff, namely a^k is his conditional expected payoff given Γ_k, and β is player 2's expected payoff.

Remarks

(1) $\Gamma(p)$ fits into the general framework of non-zero sum repeated games with lack of information on one side studied by Aumann, Maschler

and Stearns (1968), Hart (1985) and Sorin (1983) by setting

$$C^k(i,j) = C^k(j) \qquad C = A \text{ or } B, \quad i \in \{H,T\},\ j \in J.$$

(2) Except in Section 4.2, all results hold for an arbitrary number of states of nature.

(3) The payoffs in $\Gamma(p)$ are defined in such a way that the players can devote a finite number of stages to communication without affecting their long run payoff. In a typical equilibrium scenario, the phase of communication is followed by a stationary decision j of player 2 at all subsequent stages. This suggests that the game can be interpreted as a single stage decision problem (player 2's decision in J) preceded by a conversation between the two players.

3. Nash Equilibria

3.1 Non-revealing equilibria

For every $0 \le p \le 1$, $\Gamma(p)$ has a non-revealing equilibrium. Suppose player 1 plays in the same way in Γ_1 and Γ_2, i.e. plays independently of his type; then the best for player 2 is to choose an action j (or a mixture of actions) maximizing

$$b_j(p) = pB^1(j) + (1-p)B^2(j)$$

at every stage of the game. Conversely, if player 2 plays optimally in the expected game, regardless of player 1's signal, the latter can as well remain silent.

Examples of non-revealing equilibria are given in Section 3.5.

3.2 Signalling

Player 1 can reveal some information to player 2 by playing differently if Γ_1 is the state of nature or if it is Γ_2. A completely revealing strategy is for instance given by: send H if Γ_1, T if Γ_2. Obviously, this cannot always be achieved at an equilibrium; in the game

Γ_1 :

0,0	1,1

Γ_2 :

1,1	0,0

there is obviously a completely revealing equilibrium, but in the following one:

Γ_1 :	1,0	0,1
Γ_2 :	1,1	0,0

player 1 will always pretend to be of type 2. Let us derive the equilibrium conditions for signalling. A strategy for player 1 at the first stage can be described by a type-dependent lottery: if Γ_k, send H with probability q_k, T with probability $1 - q_k$, $0 \leq q_k \leq 1$, $k = 1, 2$. It has a signalling effect if $q_1 \neq q_2$. It will then induce posteriors $p_H = prob(\Gamma_1 \mid H), p_T = prob(\Gamma_1 \mid T)$. Let (a_H, β_H) (resp. (a_T, β_T)) be the payoffs if H (resp. T) is reached. For instance, player 2 makes his decision right after one stage of communication and these payoffs correspond to non-revealing equilibria (of $\Gamma(p_H)$ and $\Gamma(p_T)$ respectively). More general scenarios will be investigated below. Assume $0 < q_1 < q_2$ (then $0 < p_H < p < p_T$); if $a_H^1 < a_T^1$, say, then player 1 will not randomize between H and T in Γ_1, but will send H with probability one. Such arguments lead to the following conditions:

$$0 < q_1 < q_2 < 1, 0 < p_H < p < p_T < 1 \quad : \quad \begin{array}{l} a_H^1 = a_T^1 \\ a_H^2 = a_T^2 \end{array} \tag{3.1}$$

$$q_1 = 0 \ , \ p_H = 0 \quad : \quad a_H^1 \leq a_T^1 \tag{3.2}$$

$$q_2 = 1 \ , \ p_T = 1 \quad : \quad a_H^2 \geq a_T^2 . \tag{3.3}$$

Signalling will be used to characterize the set of all Nash equilibrium payoffs of $\Gamma(p)$; we will see that all of them can be reconstructed from the non-revealing equilibrium payoffs. Here, we have already observed that if (a_H, β_H) and (a_T, β_T) are non-revealing equilibrium payoffs of $\Gamma(p_H)$ and $\Gamma(p_T)$ respectively and $a_H = a_T = a$, then for every ρ, $0 \leq \rho \leq 1$, $(a, \rho\beta_H + (1-\rho)\beta_T)$ is an equilibrium payoff of $\Gamma(p)$ where $p = \rho p_H + (1-\rho)p_T$. Geometrically, we have "convexified in p, keeping a fixed". In order to proceed in the same fashion at all p's, including 0 and 1, we add irrelevant non-revealing payoffs at 0 and 1, so that (3.2) and (3.3) can be replaced by equalities as (3.1): at $p_H = 0$, we consider all payoffs (a^1, a_H^2) with $a^1 \geq a_H^1$ as non-revealing equilibrium payoffs and similarly at $p_T = 1$, all payoffs (a_T^1, a^2) with $a^2 \geq a_T^2$; the effect of this will become clear in Section 3.4.

Examples of signalling equilibria are given in Section 3.5.

3.3 JOINTLY CONTROLLED LOTTERIES

Let e_1 and e_2 be two equilibrium payoffs of $\Gamma(p)$ and assume the players can observe a public random variable before playing the game. Then they can achieve any convex combination of e_1 and e_2 as an equilibrium payoff.

In our model, no extraneous variable can be observed but a sequence of moves can model a conversation between the players. Let us interpret the move $j = 1$ (resp. $j = 2$) of player 2 as a message H (resp. T). The following equilibrium strategies yield the payoff $\frac{1}{2}e_1 + \frac{1}{2}e_2$: at stage 1, each player plays H with probability $\frac{1}{2}$; if the choices match (HH or TT), they go on with e_1; otherwise (HT or TH), they play e_2. Unilateral deviation is useless, hence the name "jointly controlled lottery" (see Aumann, Maschler and Stearns (1968)).

To get arbitrary convex combinations $\rho e_1 + (1 - \rho)e_2$, the players can apply the procedure repeatedly, note 0 (resp. 1) when HH or TT (resp. HT or TH) is played and interpret the sequence as the binary expansion of a uniform random variable u on $[0, 1]$. They play e_1 (resp. e_2) as soon as they see that $u \leq \rho$ (resp. $u > \rho$).

This shows that for any $0 \leq p \leq 1$, the set of equilibrium payoffs of $\Gamma(p)$ is convex. To get new equilibrium payoffs from given ones, one can thus "convexify in (a, β), keeping p fixed" (no information on the state of nature is conveyed during a jointly controlled lottery and the probability distribution p does not change).

Remark. We have just observed that with only two messages H and T, one stage of convexification at p fixed can in fact represent an unbounded number of stages of conversation; but we consider this as one single stage, as if the players could send messages in $[0, 1]$; conversations of an unbounded length will be used for deeper purposes (see below).

3.4 CHARACTERIZATION OF NASH EQUILIBRIUM PAYOFFS

We are ready to present a full description of the Nash equilibrium payoffs of $\Gamma(p)$. This is based on the results of Hart (1985), which hold in a much more general context than the present one.

First, we give some intuition for the characterization and then, we state it formally.

Let (a, β) be a Nash equilibrium payoff of $\Gamma(p)$. Suppose that in order to achieve it, the players communicate during a finite number of stages T after which player 2 makes his decision; T is a random variable, whose distribution depends on the players' strategies. Obviously, T may be deterministic; in this particular case, the length of the conversation between the players is known in advance, before it begins; but in full generality, the duration of the conversation may depend on the messages which are exchanged.

At stage T, when communication ends, player 2 has to take an action which is optimal given his information. In other words, if p_T denotes player 2's posterior probability for Γ_1 at stage T, a non-revealing equilibrium of $\Gamma(p_T)$ must be played (recall section 3.1).

How do the posteriors and the payoffs evolve from stage 1 to stage T? Let p_t be the posterior probability of Γ_1 at stage t and (a_t, β_t) be the expected payoff at stage t (i.e. given the information available at stage t). Obviously, $(p_1, a_1, \beta_1) = (p, a, \beta)$ and we have just observed that the limit point (a_T, β_T) is a non-revealing equilibrium payoff of $\Gamma(p_T)$. $p_t \neq p_{t+1}$ means that signalling (from player 1) has occured; the equilibrium conditions for player 1 require that $a_t = a_{t+1}$ (see Section 3.2). So, at every step of the process (p_t, a_t, β_t), either p_t changes in which case a_t cannot vary or p_t remains fixed and a_t can change (this happens for instance when the players make a jointly controlled lottery).

A process like (p_t, a_t, β_t) is called a *bimartingale* (Hart (1985)) : it is a martingale (the expectation of $(p_{t+1}, a_{t+1}, \beta_{t+1})$ given the past history until stage t is (p_t, a_t, β_t)) and it has a "bi-property" : for every t, either $p_{t+1} = p_t$ or $a_{t+1} = a_t$.

In general, one cannot exclude that communication phases and payoff accumulation phases alternate "ad infinitum"; in this case, the bimartingale converges to $(p_\infty, a_\infty, \beta_\infty) \in G$.

The conditions above are also sufficient for an equilibrium. Suppose there is a bimartingale (p_t, a_t, β_t) starting at (p, a, β) and ending after a finite (possibly random) time T at a non-revealing equilibrium (i.e. (a_T, β_T) is a non-revealing equilibrium payoff of $\Gamma(p_T)$). An equilibrium of $\Gamma(p)$ with payoff (a, β) is easily constructed. The process can be represented as a tree of the form

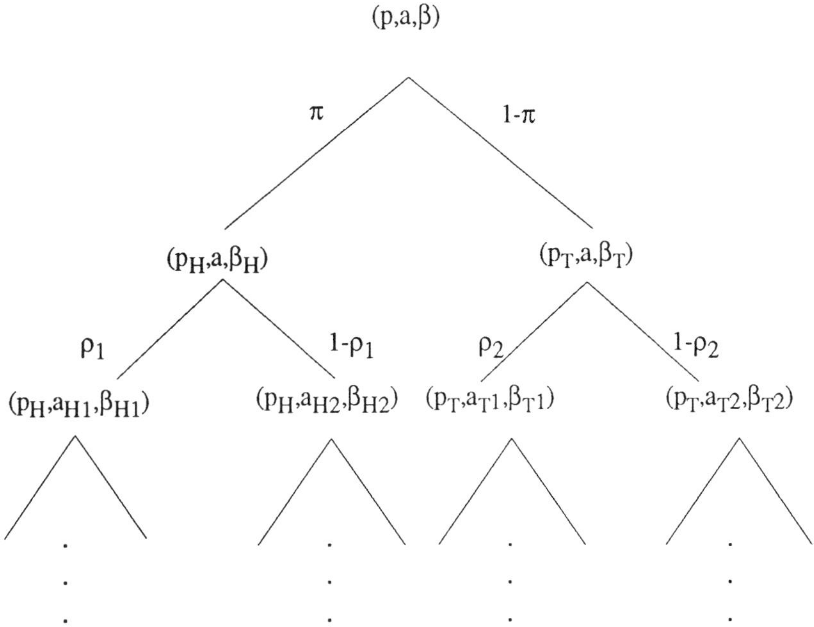

where $\pi, \rho_1, \rho_2 \in [0,1]$

$\pi(p_H, a, \beta_H) + (1-\pi)(p_T, a, \beta_T) = (p, a, \beta)$

$\rho_1(p_H, a_{H1}, \beta_{H1}) + (1-\rho_1)(p_H, a_{H2}, \beta_{H2}) = (p_H, a, \beta_H)$

etc.

The first stage will be realized by means of signalling and the second, by means of jointly controlled lotteries (recall the remark at the end of Section 3.3). This can be pursued until stage T; there, player 2 makes his decision, which is optimal since a non-revealing equilibrium is reached.

The statement below formalizes these ideas.

Definition 3.1. $(a, \beta) \in \Re^2 \times \Re$ is a non-revealing equilibrium payoff of $\Gamma(p)$ if there exists a probability distribution $\tau = (\tau_j)_{j \in J}$ on J such that

$$\beta = \sum_{j \in J} \tau_j b_j(p) = \max_{j \in J} b_j(p)$$

$$a^k = \sum_{j \in J} \tau_j A^k(j), \quad k = 1, 2$$

If $p = 0$ (resp. $p = 1$) the last condition is relaxed to $a^k \geq \sum_{j \in J} \tau_j A^k(j)$ for $k = 1$ (resp. $k = 2$).

$$G = \{(p, a, \beta) \in [0,1] \times \Re^2 \times \Re : \\ (a, \beta) \textit{ is a non-revealing equilibrium payoff of } \Gamma(p)\}.$$

Proposition 3.2. *The set of all points* (p, a, β) *where* (a, β) *is a Nash equilibrium payoff of* $\Gamma(p)$ *coincides with*

$$G^* = \{(p, a, \beta) \in [0,1] \times \Re^2 \times \Re : \text{there exists a bimartingale } (p_t, a_t, \beta_t) \text{ starting at } (p, a, \beta) \text{ and converging to } (p_\infty, a_\infty, \beta_\infty) \in G\}.$$

This result is due to Hart (1985). The geometric properties of the set G^* for an arbitrary G have been studied in Aumann and Hart (1986); G^* can be characterized by separation properties.

In the simple model considered here, one may wonder if such complex constructions are necessary. It seems that player 1 could say all he has to say at once, in one signalling stage. This is not true, as the examples of the following section will show. We will exhibit equilibrium payoffs which *require* a *random* number T of stages of conversation, i.e. which cannot be achieved if the number of stages is fixed in advance. This is the purpose of examples 3.4 and 3.5; the geometric configuration was first pointed out in Aumann and Hart (1985, example 2.5); a game was then constructed with this configuration (Forges (1984)). It is important to notice that a conversation of unbounded length would be necessary to achieve some equilibrium payoffs even if the players could send messages in [0,1] at every stage.

Observe that G is convex in (a, β) for every fixed p. Let G_1 be obtained from G by convexifying in (p, β) when a is kept constant. G_1 corresponds to the equilibrium payoffs which require one step of signalling. Now, let G_2 be constructed from a two stage convexification of G_1: first, convexification in (a, β) when p is fixed, next in (p, β) when a is fixed. By continuing in this way, one can construct an increasing sequence G_n, whose limit is the "bi-convexification" of G (denoted as Bico(G)), i.e. the smallest bi-convex (convex in a (resp. p) when p (resp. a) is fixed) set containing G. Points in Bico(G) correspond to equilibrium payoffs which can be obtained with a conversation of deterministic length. The examples of the next section will illustrate that G_1 need not coincide with G_2 and that Bico(G) may be different from G^*.

3.5 Examples

Example 3.3. We use the notation $a_j = (A^1(j), A^2(j)), j \in J$.

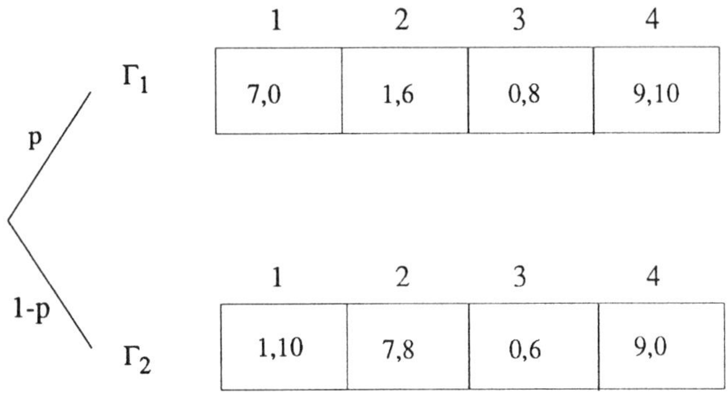

First, let us get a description of the non-revealing equilibria. Player 2 maximizes his expected payoff:

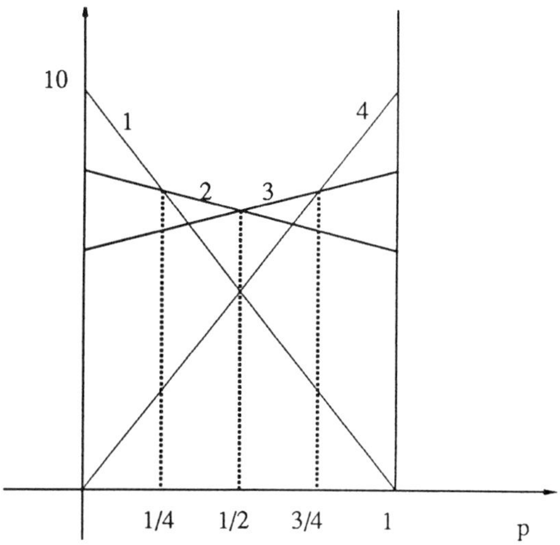

hence, he plays action 1 if $0 \le p \le \frac{1}{4}$, 2 if $\frac{1}{4} \le p \le \frac{1}{2}$, 3 if $\frac{1}{2} \le p \le \frac{3}{4}$, 4 if $\frac{3}{4} \le p \le 1$. In particular, at the values of p where two actions are optimal, he can use any mixture of the two. Let us represent the corresponding non revealing equilibrium payoffs (a^1, a^2) of player 1:

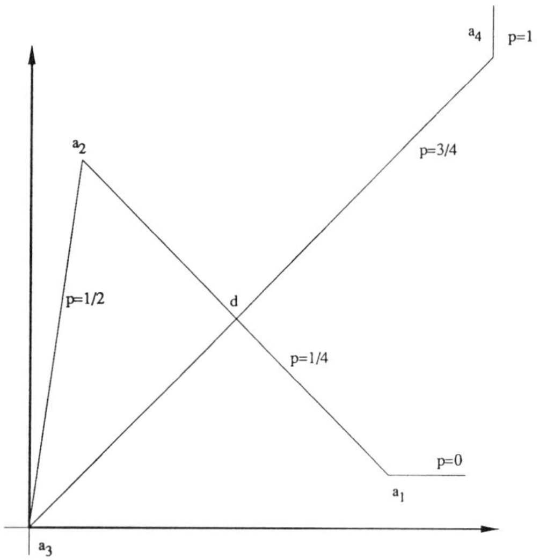

Figure 1

G can be described as follows:

interval for p	corresponding extreme a payoffs of player 1	expected payoff β of player 2
0	$a_1 (a_1^1 \geq 1)$	10
$(0, \frac{1}{4})$	a_1	$10(1-p)$
$\frac{1}{4}$	$a_1 a_2$	7.5
$(\frac{1}{4}, \frac{1}{2})$	a_2	$6p + 8(1-p)$
$\frac{1}{2}$	$a_2 a_3$	7
$(\frac{1}{2}, \frac{3}{4})$	a_3	$8p + 6(1-p)$
$\frac{3}{4}$	$a_3 a_4$	7.5
$(\frac{3}{4}, 1)$	a_4	$10p$
1	$a_4 (a_4^2 \geq 9)$	10

Let us investigate the opportunities for *signalling*. Complete revelation of the game cannot be achieved at an equilibrium. If player 2 knew the "true game", he would play 1 in Γ_2 and 4 in Γ_1. But as the two types of player 1 prefer 4 to 1, type 2 would lie and pretend to be type 1. Geometrically, this is expressed by the fact that the segments from a_1 and a_4, parallel to the axes, do not intersect.

¿From our preceding analysis, in order to get signalling, we must have two non-revealing equilibria, at two different p's, giving the same payoff to player 1. The following points of G : $(\frac{1}{4}, (4,4), 7.5)$ and $(\frac{3}{4}, (4,4), 7.5)$ satisfy this requirement (the points are easily found on Figure 1 where $d = (4,4)$; here the two equilibria give also the same payoff to player 2, which is purely incidental). Let $\frac{1}{4} < p < \frac{3}{4}$, for instance $p = \frac{1}{2}$; to get $((4,4), 7.5)$ at $p = \frac{1}{2}$, the players proceed as follows : if he is of type 1 (resp. type 2), player 1 sends H with probability $\frac{1}{4}$ (resp. $\frac{3}{4}$) and T with probability $\frac{3}{4}$ (resp. $\frac{1}{4}$), in order to generate the posteriors $p_H = \frac{1}{4}$ and $p_T = \frac{3}{4}$. (One can think of player 1 using two biased coins). Player 2 reacts as follows : if he receives the message H (resp. T) he plays 1 with probability $\frac{1}{2}$ and 2 with probability $\frac{1}{2}$ (resp. 3 with probability $\frac{5}{9}$ and 4 with probability $\frac{4}{9}$).

By proceeding in a similar way for every $\frac{1}{4} < p < \frac{3}{4}$ and by convexifying then in the payoffs when p is fixed, we get the following description for the projection of G_1 on the (p, a)-coordinates.

interval for p	corresponding extreme payoffs a of player 1
0	$a_1 (a_1^1 \geq 1)$
$(0, \frac{1}{4})$	a_1
$\frac{1}{4}$	$a_1 a_2$
$(\frac{1}{4}, \frac{1}{2})$	$a_2 d$
$\frac{1}{2}$	$a_2 a_3 d$
$(\frac{1}{2}, \frac{3}{4})$	$a_3 d$
$\frac{3}{4}$	$a_3 a_4$
$(\frac{3}{4}, 1)$	a_4
1	$a_4 (a_4^2 \geq 9)$

In particular, at $p = \frac{1}{2}$, we have now the whole triangle a_2 a_3 d. One does not get new equilibrium payoffs by repeating the procedure.

The next examples are reprinted from Forges (1984) (with the permission of the International Journal of Game Theory). As announced in section 3.4, they illustrate that one may have $G_1 \subsetneq G_2$ or even $Bico(G) \subsetneq G^*$. More precisely, they show that $Bico(G) \subsetneq G^{\#}$ where $G^{\#}$ is the subset of G^* obtained by requiring that the bimartingale converges in a finite, possibly random, time; it can be checked that in the examples below, $G^{\#} = G^*$ (hence, the reader may ignore the difference between $G^{\#}$ and G^* here). The geometric configuration appearing in example 3.4 was found by Aumann and Hart ((1986), example 2.5, known as the "four frogs") independently of any underlying game. Example 3.4 provides a game where it indeed arises.

Example 3.4. Let $\Gamma(p)$ be defined as in the first example, the payoff matrices of Γ_1 and Γ_2 being now

	1	2	3	4	5	6	7	8	9
Γ_1 :	6,-6	2,1	8,6	4,9	0,-10	0,-2	0,4	0,8	0,10

	1	2	3	4	5	6	7	8	9
Γ_2 :	4,9	8,6	2,1	6,-6	0,10	0,8	0,4	0,-2	0,-10

The projection of the space G on $[0,1] \times \Re^2$ is given by

$$\{(p,0) \mid p \in [0,1]\} \cup \{(j/5, a) \mid a \in [0, a_j], j = 1, \cdots, 4\}$$

using the same notations as before. One checks immediately that this set is bi-convex. It has the following intersection with the plane $a^1 + a^2 = 10$.

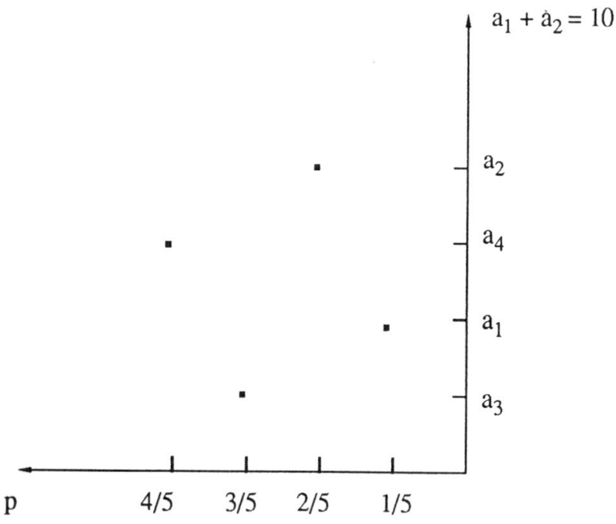

which is exactly the configuration of the "four frogs". The projection of G^* on the same plane is

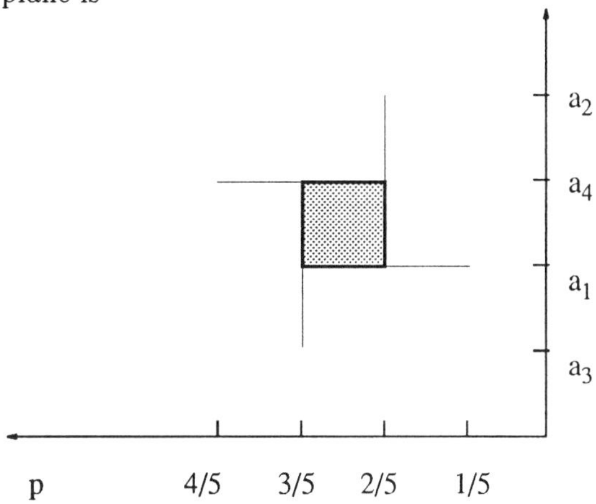

Indeed, let us consider an arbitrary point of the square above. Let us show how to construct a bimartingale starting at this point and converging to a point of G. First, one goes to the boundary of the square (in a parallel direction with one of the axes); being on a side, one moves to its extreme points, stopping as soon as a point of G is reached. It is quite clear that the number of stages needed to attain an element of G is not bounded. Now, for $p \in (2/5, 3/5)$, 0 is the only nonrevealing equilibrium payoff for player 1. He improves thus considerably his payoff by achieving a point of G^*.

The game considered here is very tractable but rather degenerate; the example may nevertheless be used to construct completely generic games with the same property. Consider for instance the game $\Gamma(p)$ defined by

	1	2	3	4	5
Γ_1 :	0,-5	0,-1	-1,2	3,4	1,5

	1	2	3	4	5
Γ_2 :	1,5	3,4	-1,2	0,-1	0,-5

Example 3.5. In this example, we have $G \subsetneqq G_1 \subsetneqq G_2 = Bico(G) \subsetneqq G^*$. To check this, we can proceed as for the preceding examples. First, let us represent the vector payoffs a_j of player 1 when player 2 uses his pure strategy $j \quad (j = 1, \cdots, 5)$.

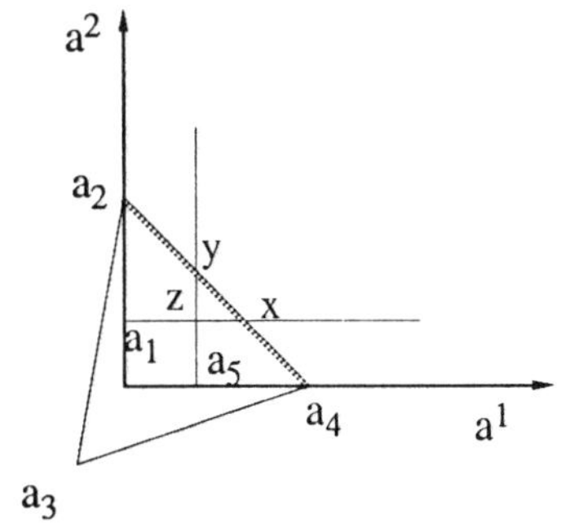

Using again the results of Hart (1985), we get that the projection of G on $[0,1] \times \Re^2$ is the convex hull in "a" of

$$\left\{ (p,a) \;\middle|\; \begin{array}{l} p=0,\ a^1>0,\ a^2=1 \\ p \in \left[\frac{j-1}{5}, \frac{j}{5}\right],\ a=a_j, \quad j=1,\ldots,5 \\ p=1,\ a^1=1,\ a^2>0\ . \end{array} \right\}$$

Applying the same constructions as in the first example, we get G_1 from G by adding the vector payoff z (see the graph above) at every $p \in (0,1)$. Note that this payoff corresponds to an equilibrium where player 1 reveals the true game. Convexifying then G_1 when a is fixed, we obtain the following description of G_2:

interval for p	corresponding extreme payoffs of player 1
0	$a_1 x$
$(0, \frac{1}{5})$	$a_1 z$
$[\frac{1}{5}, \frac{2}{5})$	$a_1 a_2 z$
$\frac{2}{5}$	$a_2 a_3 z$
$(\frac{2}{5}, \frac{3}{5})$	$a_3 z$
$\frac{3}{5}$	$a_3 a_4 z$
$(\frac{3}{5}, \frac{4}{5}]$	$a_4 a_5 z$
$(\frac{4}{5}, 1)$	$a_5 z$
1	$a_5 y$

It appears thus that $G_2 = Bico(G)$. In the plane $a^1 + a^2 = 3$, we have the following projection

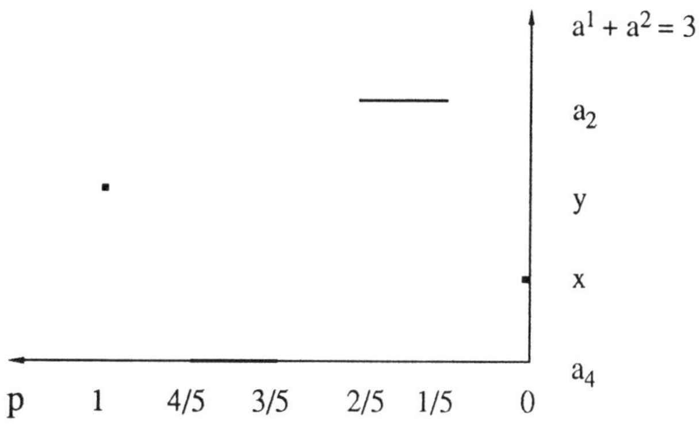

As in the second example, $Bico(G) \subsetneq G^*$, the projection of the latter space on the same plane is indeed

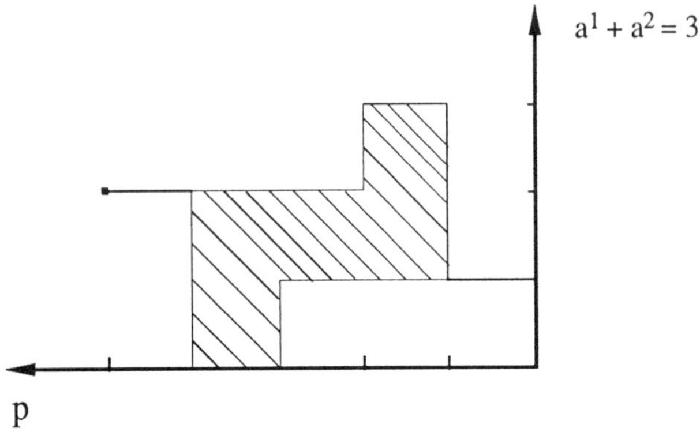

For instance, at $p = \frac{1}{2}$, the payoff $a = \left(\frac{3}{2}, \frac{3}{2}\right)$ can be achieved. Let us compute the corresponding expected payoff β of player 2. A bimartingale starts from $\left(\frac{1}{2}, \left(\frac{3}{2}, \frac{3}{2}\right), \beta\right)$ and reaches G; without loss of generality, by adding a further splitting if necessary, one can assume that it converges to

$$\begin{array}{lll} (0,(2,1),5) & (\frac{1}{5},(0,3),3) & (\frac{2}{5},(0,3),2) \\ (1,(1,2),5) & (\frac{4}{5},(3,0),3) & (\frac{3}{5},(3,0),2) \end{array}.$$

The corresponding expected payoffs $pa^1 + (1-p)a^2$ of player 1 are $1, \frac{12}{5}, \frac{9}{5}$. By the martingale property of the expected payoffs, there must be weights $\delta_1, \delta_2, \delta_3$ such that

$$\delta_1 \cdot 1 + \delta_2 \cdot \frac{12}{5} + \delta_3 \cdot \frac{9}{5} = \frac{3}{2}$$

$$\delta_1 \cdot 5 + \delta_2 \cdot 3 + \delta_3 \cdot 2 = \beta$$

so that β will vary from $\frac{25}{8}$ to $\frac{30}{7}$ (the first bound corresponds to $\delta_1 = \frac{3}{8}, \delta_3 = \frac{5}{8}$, the second, to $\delta_1 = \frac{9}{14}, \delta_2 = \frac{5}{14}$).

4. Correlated Equilibria

4.1 Definition and characterization

According to the definition introduced by Aumann (1974,1987), a *correlated equilibrium* in $\Gamma(p)$ is a Nash equilibrium of an extension of $\Gamma(p)$ where before the beginning of the game, a correlation device selects a pair of signals, one for each player. The signals may be correlated and the players may base their strategies in $\Gamma(p)$ on their signals; hence, the correlated equilibrium solution concept is appropriate to model coordination between the players.

At first sight, correlated equilibria look more complex than Nash equilibria. We have just seen that a Nash equilibrium of $\Gamma(p)$ may use a subtle process of information transmision. If we must also capture coordination effects, we expect a more complex analysis.

This intuition is wrong. The characterization of correlated equilibria is much more tractable than the one of Nash equilibria. To state it, we need some terminology.

Let us call *noisy channel* a device described by a couple (λ, μ) of probability distributions on J and used as follows: player 1 is asked to reveal his type to the noisy channel; if player 1 announces Γ_1 (resp. Γ_2), the probability distribution λ (resp. μ) is used to choose an action $j \in J$, which is then suggested to player 2.

At a *noisy channel equilibrium*, player 1 says the truth and player 2 follows the recommendation.

Observe that a noisy channel preserves the non-cooperative structure of the game; in particular, an advice is given to player 2 but he has to decide himself. Observe also that if a noisy channel is added to $\Gamma(p)$, the conversation between the players is not necessary (even more generally, any communication equilibrium payoff of $\Gamma(p)$ can be obtained by means of a noisy channel, see Forges (1985, 1986)).

This shows an important difference between a correlated equilibrium and a noisy channel equilibrium: in the first case, player 1 communicates directly with player 2, while in the second case, an informational intermediary is used. Nevertheless, the following characterization holds:

Proposition 4.1. *The set of correlated equilibrium payoffs of* $\Gamma(p)$ *coincides with the set of noisy channel equilibrium payoffs.*

This result is established in Forges (1985), for an arbitrary number of states of nature. The proof shows an additional interesting property: for every noisy channel equilibrium, one can construct a payoff-equivalent correlated equilibrium, which can be achieved with *one stage* of information transmission from player 1 to player 2; a sufficiently large (but *finite*) alphabet of messages is necessary. Correlated equilibria behave thus completely differently from Nash equilibria.

4.2 Geometric properties

It is interesting to see how the characterization above can be used to solve specific games or to construct examples with given features.

Let us denote as $Q(p)$ the set of all couples (λ, μ) of probability distributions on J determining a noisy channel equilibrium of $\Gamma(p)$. The linear inequalities to be satisfied by (λ, μ) are easily derived. For player 1, they express that if the state of nature is Γ_1 (resp. Γ_2), he cannot gain by pretending it is Γ_2 (resp. Γ_1). Suppose action j is recommended to player 2, he computes a posterior probability p_j given by

$$p_j = \frac{\lambda_j p}{\lambda_j p + \mu_j(1-p)}.$$

He will be in equilibrium if action j is optimal at p_j in the sense that action j maximizes his expected payoff $p_j B^1(i) + (1-p_j)B^2(i), i \in J$. Now, the unit interval can be partitioned into subintervals $\Delta_j = [d_j^-, d_j^+]$ where action j is optimal, so that the equilibrium conditions for player 2 are of the form $d_j^- \le p_j \le d_j^+$. Hence $(\lambda, \mu) \in Q(p)$ if and only if

$$\begin{aligned} &\lambda_j \ge 0 \quad \mu_j \ge 0 \qquad \forall\, j \in J \\ &\sum_{j\in J} \lambda_j = \sum_{j \in J} \mu_j = 1 \end{aligned} \tag{4.1}$$

$$\begin{aligned} &\sum_{j\in J} (\lambda_j - \mu_j) A^1(j) \ge 0 \\ &\sum_{j\in J} (\lambda_j - \mu_j) A^2(j) \le 0 \end{aligned} \tag{4.2}$$

$$\begin{aligned} &\lambda_j p(1-d_j^-) - \mu_j d_j^-(1-p) \ge 0 \qquad \forall\, j \in J \\ &\lambda_j p(1-d_j^+) - \mu_j d_j^+(1-p) \le 0 \qquad \forall\, j \in J. \end{aligned} \tag{4.3}$$

The following proposition states that the posterior probability distributions corresponding to an extreme point (λ, μ) of $Q(p)$ cannot coincide with p with positive probability unless this probability is one (in which case the noisy channel equilibrium is non-revealing, containing but one signal) and that otherwise these posterior probability distributions necessarily lie at the boundary of Δ_j. As we shall see later, this will provide a complete description of $Q(p)$.

Proposition 4.2. *Let (λ, μ) be an extreme point of $Q(p)$; for every $j \in J$ such that $(\lambda_j, \mu_j) \neq (0,0)$,*

(a) $p_j = p \Rightarrow \lambda_j = \mu_j = 1$

(b) *otherwise,* $p_j = d_j^-$ *or* d_j^+.

Proof:

(a) If $p_j = p$, then $\lambda_j = \mu_j$. Assume this number is not equal to 1; then (λ, μ) will be written as a convex combination of two points $(\lambda^1, \mu^1), (\lambda^2, \mu^2)$ of $Q(p)$. (λ^1, μ^1) corresponds to staying at p with probability one and playing j there:

$$\begin{aligned} \lambda_j^1 &= \mu_j^1 = 1 \\ \lambda_i^1 &= \mu_i^1 = 0 \qquad \text{if } i \neq j \end{aligned}$$

(λ^2, μ^2) is associated with the other posterior probabilities:

$$\begin{aligned} &\lambda_j^2 = \mu_j^2 = 0 \\ &\lambda_i^2 = \frac{\lambda_i}{1-\lambda_j} \qquad \mu_i^2 = \frac{\mu_i}{1-\mu_j} = \frac{\mu_i}{1-\lambda_j} \quad \text{if } i \neq j. \end{aligned}$$

It is easily checked that these points belong to $Q(p)$ and

$$(\lambda, \mu) = \lambda_j(\lambda^1, \mu^1) + (1-\lambda_j)(\lambda^2, \mu^2).$$

(b) Let now $p_j \neq p$. Suppose that p_j is an interior point of Δ_j i.e. that the inequalities (4.3) are strict. In particular, $0 < \lambda_j < 1$ and $0 < \mu_j < 1$. As above, we construct two points $(\lambda^1, \mu^1), (\lambda^2, \mu^2)$ of $Q(p)$ of which (λ, μ) is a convex combination. Set

$$\begin{aligned} \lambda_j^1 &= \frac{\lambda_j - \varepsilon}{1-\varepsilon} \qquad & \lambda_j^2 &= \frac{\lambda_j + \varepsilon}{1+\varepsilon} \\ \mu_j^1 &= \frac{\mu_j - \varepsilon}{1-\varepsilon} \qquad & \mu_j^2 &= \frac{\mu_j + \varepsilon}{1+\varepsilon} \end{aligned}$$

with $0 < \varepsilon < \min\{\lambda_j, \mu_j\}$; ε can be chosen so as to satisfy further

$$\begin{aligned} &\varepsilon|p(1-d_j^-) - d_j^-(1-p)| < \lambda_j p(1-d_j^-) - \mu_j d_j^-(1-p) \\ &\varepsilon|p(1-d_j^+) - d_j^+(1-p)| < -[\lambda_j p(1-d_j^+) - \mu_j d_j^+(1-p)] \end{aligned}$$

i.e. so that (4.3) holds for (λ_j^1, μ_j^1) and (λ_j^2, μ_j^2). We complete the definition of $(\lambda^1, \mu^1), (\lambda^2, \mu^2)$ by

$$\begin{aligned} \lambda_i^1 &= \frac{\lambda_i}{1-\varepsilon} \qquad & \lambda_i^2 &= \frac{\lambda_i}{1+\varepsilon} \\ \mu_i^1 &= \frac{\mu_i}{1-\varepsilon} \qquad & \mu_i^2 &= \frac{\mu_i}{1+\varepsilon} \end{aligned}$$

for $i \neq j$. It is easily checked that these points belong to $Q(p)$ and

$$(\lambda, \mu) = \frac{1-\varepsilon}{2}(\lambda^1, \mu^1) + \frac{1+\varepsilon}{2}(\lambda^2, \mu^2).$$

∎

Remark. If $(\lambda, \mu) \in Q(p)$ is such that $\lambda_j = \mu_j = 1$ then it is an extreme point; it corresponds to a non-revealing equilibrium.

Corollary 4.3. *Let (λ, μ) be an extreme point of $Q(p)$; at most 4 actions $j \in J$ of player 2 have positive probability (i.e. are such that $(\lambda_j, \mu_j) \neq (0,0)$).*

Proof: Suppose exactly $n \geq 2$ actions of player 2 have positive probability under (λ, μ). Given (4.1), at least $2(n-1)$ constraints among (4.2) and (4.3) must be active at (λ, μ). Since $n \geq 2$, we are in case (b) of the proposition and (λ, μ) satisfies n equalities of the form $p_j = d_j^-$ or $p_j = d_j^+$. No other constraints can be active among (4.3). Since there are two conditions for player 1 - (4.2) -, at least $n+2$ constraints will hold with an equality. Hence, $n + 2 \geq 2(n-1)$ or, $n \leq 4$. ∎

In order to derive a full characterization of the extreme points of $Q(p)$, there remains to show how to verify if $\Gamma(p)$ has a noisy channel equilibrium where each action $j \in J'$, for some subset J' of J, has positive probability. By the previous analysis, we need only to consider $2 \leq |J'| \leq 4$ and we can assume $p_j \neq p$ (obviously, $p_j \in \Delta_j$). Set

$$J_- = \{j \in J' : p_j < p\}$$
$$J_+ = \{j \in J' : p_j > p\}.$$

Assume that there exists a noisy channel $((\lambda_j)_{j \in J'}, (\mu_j)_{j \in J'})$ at p yielding the posterior probabilities p_j. Then $\lambda_j - \mu_j < 0$ (resp. > 0) for $j \in J_-$ (resp. J_+). Let us choose $\rho > 0$ such that

$$r_j = \rho(\mu_j - \lambda_j) \quad \text{for} \quad j \in J_-$$
$$s_j = \rho(\lambda_j - \mu_j) \quad \text{for} \quad j \in J_+$$

define strategies $r = (r_j)$ and $s = (s_j)$ with respective supports J_- and J_+. Recalling inequalities (4.2), we see that a necessary and sufficient condition for the noisy channel above to be in equilibrium is that there exist convex combinations $A_s = (A_s^1, A_s^2)$ and $A_r = (A_r^1, A_r^2)$ of the payoffs corresponding to the strategies r and s respectively such that $A_s^1 \geq A_r^1$ and $A_s^2 \leq A_r^2$. One can also check on player 1's vector payoffs if one or two of these inequalities can be an equality ($A_s^1 = A_r^1$ or $A_s^2 = A_r^2$), which is useful to find the extreme points of $Q(p)$, as we will see now.

To obtain a full description of the noisy channel equilibria of $\Gamma(p)$, one can thus proceed as follows. First, one determines the non-revealing equilibria (in particular, the intervals Δ_j); the corresponding noisy channels have the form $\lambda_j = \mu_j = 1$ and $p_j = p$. Then, by considering the configuration of player 1's vector payoffs $a_j = (A^1(j), A^2(j)), j \in J$, one can see for every subset J' of J of size n, $2 \leq n \leq 4$ and every possible J_- and J_+ if there is a noisy channel equilibrium where the actions in J' have positive probability. For $n = 2$, the extreme points are determined by solving two equations of the form $p_j = d_j^-$ or $p_j = d_j^+$. For $n = 3$, there are three similar linearly independent equations; one can check on player 1's payoffs if one of his constraints can be active; if it is the case, this gives a fourth equation to determine the extreme noisy channel. Finally, for $n = 4$, four linearly independent equations result from the fact that the posteriors p_j are extreme points of Δ_j. On the representation of player 1's payoffs, one can see whether his two constraints can hold with an equality, in which case two additional equations are available to determine the extreme noisy channel.

Let us consider some examples. We assume that player 2's actions have been numbered so that the subintervals Δ_j of $[0, 1]$ follow each other, in the order $\Delta_1, \Delta_2, \cdots$. Here are a few configurations for three signals .

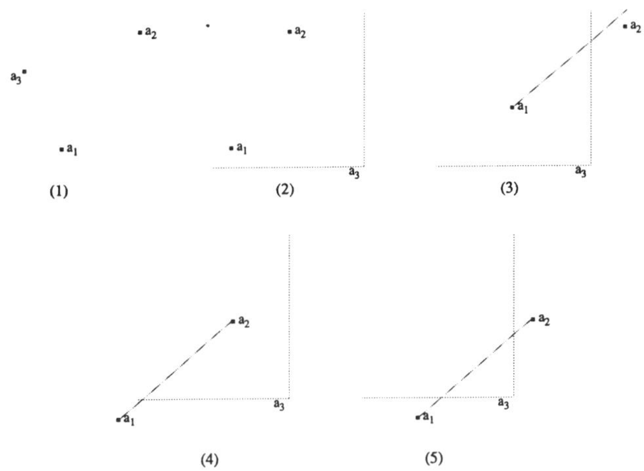

In (1), all correlated equilibria are non-revealing. In (2), there are extreme noisy channels with two signals : with $J_+ = \{3\}$ and $J_- =$ either $\{1\}$ or $\{2\}$. There is also a noisy channel with three signals ($J_- = \{1, 2\}, J_+ = \{3\}$) but is is not extreme, since player 1's constraint cannot hold with an equality; this noisy channel is indeed a convex combination of the preceding ones. In (3), we can have 2 signals with $J_- = \{1\}, J_+ = \{3\}$ or 3 signals with $J_- = \{1, 2\}, J_+ = \{3\}$. In this case, the second constraint for player 1 can hold with an equality. (4) is similar with either $J_- = \{2\}, J_+ = \{3\}$ or $J_- = \{1, 2\}, J_+ = \{3\}$ and the first constraint of player 1. In (5), there are no noisy channel equilibria with 2 signals but two kinds of extreme ones with $J_- = \{1, 2\}, J_+ = \{3\}$. Obviously, these noisy channels are only realizable at those p's where the corresponding sets J_-, J_+, make sense. Other configurations will be observed in the examples which follow.

4.3 EXAMPLES

Example 4.4. An interesting example of correlated equilibrium can be found in the last game considered in the section on Nash equilibria (example 3.5). A very complex equilibrium of $\Gamma(\frac{1}{2})$ has been constructed there with payoff $((\frac{3}{2}, \frac{3}{2}), \frac{30}{7})$. Consider now the following noisy channel. In our previous notation, $\lambda_2 = \mu_4 = \frac{1}{5}, \lambda_4 = \mu_2 = \frac{4}{5}$, the other components are 0. It yields an equilibrium; in particular, $p_2 = \frac{1}{5}$ and $p_4 = \frac{4}{5}$ so that the noisy channel is extreme. The associated payoff is $((\frac{12}{5}, \frac{12}{5}), 3)$.

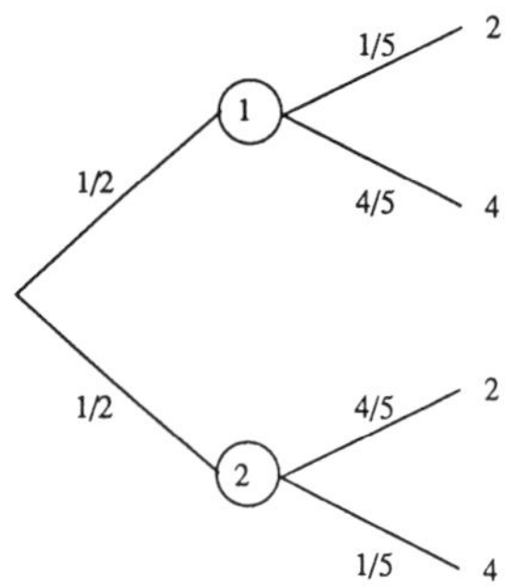

This is the best correlated equilibrium for player 1 (see Figure 2). Moreover, it cannot be achieved as a Nash equilibrium (to prove this, one can use the geometric characterization of Aumann and Hart (1986)). Using the system of linear inequalities (4.1) to (4.3) one easily gets a description of all expected payoffs $\alpha = \frac{1}{2}(a^1 + a^2)$, β for $p = \frac{1}{2}$ (see Figure 2).

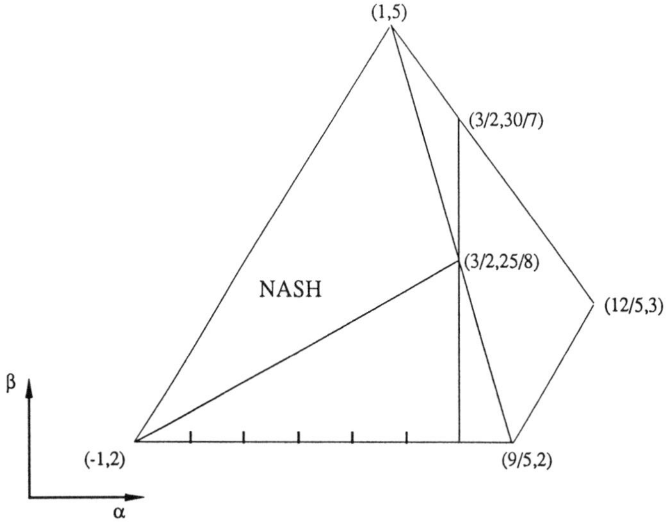

Figure 2

One sees that the Nash equilibrium payoff $(\frac{3}{2}, \frac{30}{7})$ which requires that no deadline be imposed to the conversation of the players can be achieved as a very simple correlated equilibrium payoff : it is a convex combination of the equilibrium above and the completely revealing equilibrium, with respective weights $\frac{9}{14}, \frac{5}{14}$. The corresponding noisy channel is thus $(\lambda, \mu) = \frac{1}{14}((0,1,4,9),(9,4,1,0))$.

One can construct an equivalent correlated equilibrium where player 1 himself will transmit information to player 2, without making inputs to a machine (by using the procedure of Forges (1985)).

Example 4.5. Let us now go back to example 3.3. Player 1's payoffs configuration (Figure 1) shows that at $p = \frac{1}{2}$, one has an extreme noisy channel with $J_- = \{1,2\}$ and $J_+ = \{3,4\}$. If we take $p_1 = p_2 = \frac{1}{4}$ and $p_3 = p_4 = \frac{3}{4}$, we just get the Nash equilibrium payoff we have already discovered. But if we take $p_1 = 0, p_2 = \frac{1}{4}, p_3 = \frac{3}{4}, p_4 = 1$, we get the noisy channel $(\lambda, \mu) = \frac{1}{55}((0,9,30,16),(18,27,10,0))$ (besides the conditions on the posteriors, we have imposed that the two constraints of player 1 be active, as explained in the previous section). The associated payoff is $((\frac{153}{55}, \frac{207}{55}), \frac{455}{55})$; this improves player 2's expected payoff from Nash equilibria: then, it could not be greater than 7.5.

Example 4.6. Let us keep the same payoffs for player 2 as in the preceding example, but take another configuration for player 1's payoffs.

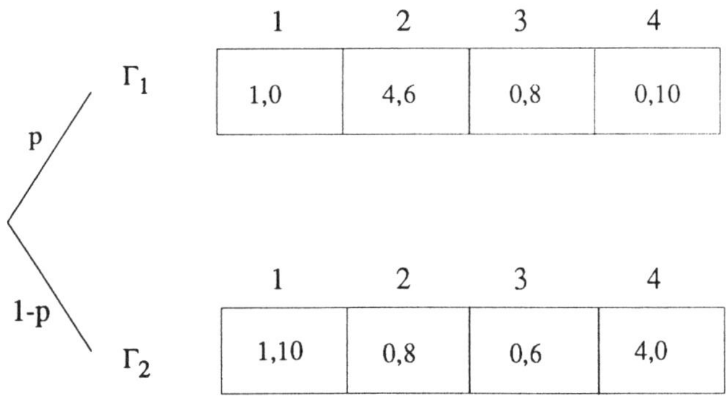

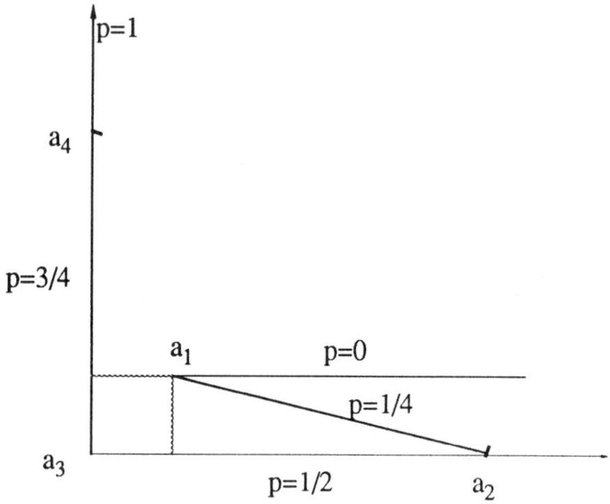

The Nash equilibria are all non-revealing; but player 1's payoffs show that there is an extreme noisy channel with four signals and $J_- = \{1\}, J_+ = \{2,3,4\}$; this is feasible for $0 < p < \frac{1}{2}$. This will provide correlated equilibrium payoffs for player 1 which lie outside the convex hull of his Nash equilibrium payoffs, even though his equilibrium conditions hold with an equality in the noisy channel representation. Take for instance $p = \frac{1}{8}$ and consider the extreme noisy channel for which

$p_1 = 0, p_2 = \frac{1}{4}, p_3 = p_4 = \frac{3}{4}$: $(\lambda, \mu) = \frac{1}{98}((0,35,42,21),(80,15,2,1))$; player 1's vector payoff is $(\frac{10}{7}, \frac{6}{7}) \neq (1,1) = a_1$.

We finish with a complete description of the correlated equilibrium payoffs of a game with three actions for player 2 (this is reprinted from Forges (1985) with the permission of the International Journal of Game Theory).

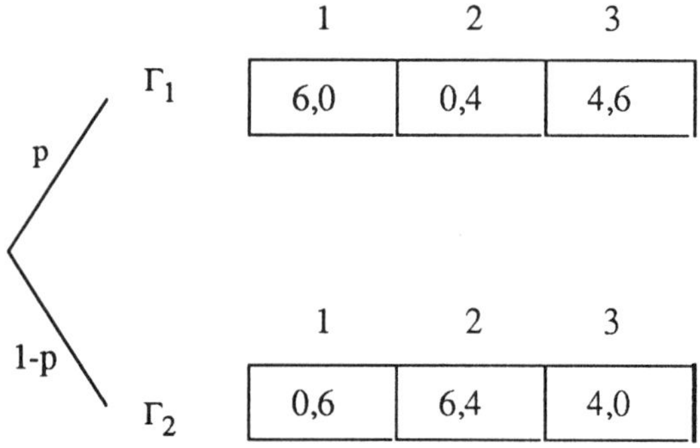

Example 4.7.

We have here $\Delta_1 = [0,1/3], \Delta_2 = [1/3,2/3], \Delta_3 = [2/3,1]$. The vector payoffs of player 1 corresponding to the various pure strategies of player 2 can be represented as follows:

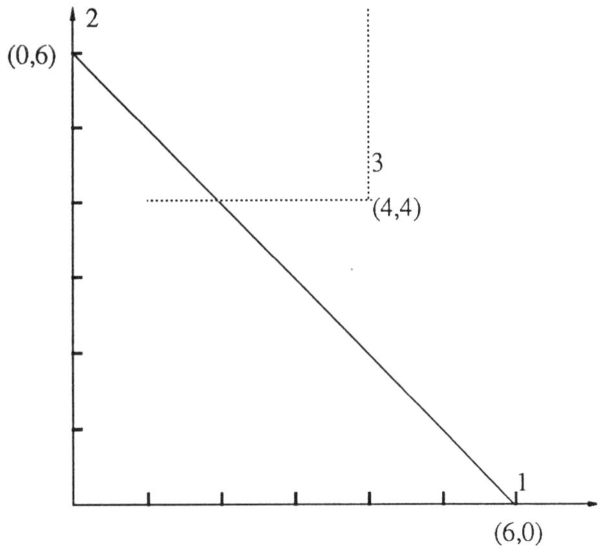

It appears that the admissible noisy channels including two signals involve necessarily the pure strategies 2 and 3. A noisy channel including the three strategies yields an equilibrium if and only if $J_- = \{1, 2\}, J_+ = \{3\}$ (i.e. $p_1 \leq p_2 < p < p_3$) and we shall get extreme points there when the second condition of player 1 is satisfied with an equality. We are now ready to compute all the extreme points of $Q(p)$, p ranging over $(0, 1)$. In the table below, we have computed (as a function of $p, p \in (\frac{1}{3}, 1)$) the expected payoffs of player 1 and player 2 for every extreme point of $Q(p)$ including two or three signals.

	posterior probabilities			admissible interval for p	expected payoff of player $1(\alpha)$	expected payoff of player $2(\beta)$
	p_1	p_2	p_3			
1	-	$\frac{1}{3}$	$\frac{2}{3}$	$(\frac{1}{3}, \frac{2}{3})$	4	4
2	-	$\frac{1}{3}$	1	$(\frac{1}{3}, 1)$	4	$3(p+1)$
3	-	$\frac{2}{3}$	1	$(\frac{2}{3}, 1)$	$2(3p-1)$	$6p$
4	0	$\frac{1}{3}$	$\frac{2}{3}$	$(\frac{1}{3}, \frac{2}{3})$	$\frac{6p(3p+1)}{6p-1}$	$\frac{3(-3p^2+11p-2)}{6p-1}$
5	0	$\frac{1}{3}$	1	$(\frac{1}{3}, 1)$	$\frac{12p(p+1)}{7p-1}$	$\frac{6(2p^2+5p-1)}{7p-1}$
6	0	$\frac{2}{3}$	1	$(\frac{2}{3}, 1)$	$\frac{12p(2p-1)}{5p-2}$	$\frac{6(2p-1)(p+2)}{5p-2}$
7	$\frac{1}{3}$	$\frac{1}{3}$	$\frac{2}{3}$	$(\frac{1}{3}, \frac{2}{3})$	$\frac{2}{3}(3p+4)$	4
8	$\frac{1}{3}$	$\frac{1}{3}$	1	$(\frac{1}{3}, 1)$	$p+3$	$3(p+1)$
9	$\frac{1}{3}$	$\frac{2}{3}$	1	$(\frac{2}{3}, 1)$	$\frac{p(9p-5)}{2p-1}$	$\frac{9p^2-p-2}{2p-1}$

To this, we must add the expected payoffs corresponding to the nonrevealing equilibria, namely

$$(6(1-p),4) \quad for \quad \frac{1}{3} < p \leq \frac{2}{3}$$
$$(4,6p) \quad for \quad \frac{2}{3} \leq p < 1 \ .$$

At $p = \frac{1}{2}$, one has the following description:

	vector payoff of player 1	expected payoff of player 1	expected payoff of player 2
N.R	$(0, 6)$	3	4
1	$(\frac{8}{3}, \frac{16}{3})$	4	4
2	$(2, 6)$	4	$\frac{9}{2}$
4	$(3, \frac{9}{2})$	$\frac{15}{4}$	$\frac{33}{8}$
5	$(\frac{12}{5}, \frac{24}{5})$	$\frac{18}{5}$	$\frac{24}{5}$
7	$(\frac{10}{3}, 4)$	$\frac{11}{3}$	4
8	$(3, 4)$	$\frac{7}{2}$	$\frac{9}{2}$

This yields the following projections on the space of vector payoffs of player 1 and on the one of expected payoffs respectively:

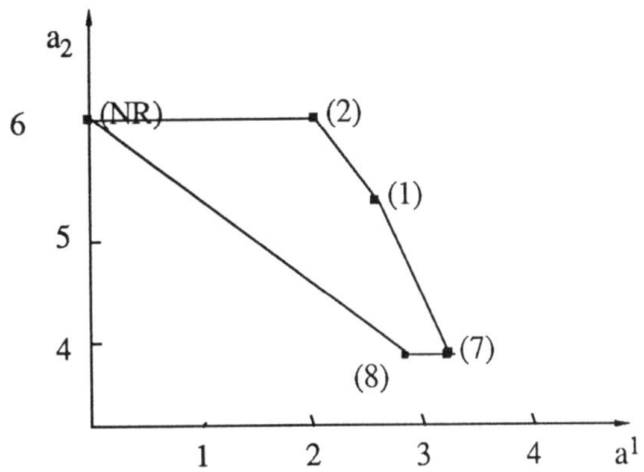

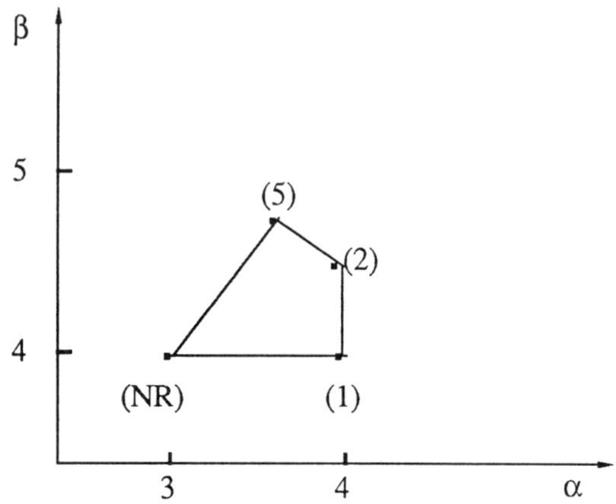

If we consider the set of equilibrium payoffs (a^1, a^2, β) at $p = \frac{1}{2}$, we remark that all the extreme noisy channels except (4) lead to extreme equilibrium payoffs. The equilibrium payoff $(3, \frac{9}{2}, \frac{33}{8})$ of (4) can be written as a convex combination of the payoffs associated with (2) and (7) (with respective weights $\frac{1}{4}$ and $\frac{3}{4}$).

5. References

[1] R. J. Aumann, "Subjectivity and correlation in randomized strategies," *Journal of Mathematical Economics* **1** (1974) 67–95.

[2] R. J. Aumann, "Correlated equilibrium as an expression of Bayesian rationality," *Econometrica* **55** (1987) 1–18.

[3] R. J. Aumann and S. Hart, "Bi-convexity and bi-martingales, *Israel Journal of Mathematics* **54** (1986) 159–180.

[4] R. J. Aumann and M. Maschler, "Game-theoretic aspects of gradual disarmament," *Mathematica*, ST-80, 1966, Ch. V, 1–55.

[5] R. J. Aumann and M. Maschler, "Repeated games with incomplete information: a survey of recent results," *Mathematica*, ST-116, 1967, Ch. III, 287–403.

[6] R. J. Aumann and M. Maschler, "Repeated games of incomplete information: the zero-sum extensive case," *Mathematica*, ST-143, 1968, Ch. V, 37–116.

[7] R. J. Aumann, M. Maschler, and R. Stearns, "Repeated games of incomplete information: an approach to the non-zero sum case," *Mathematica*, ST-143, 1968, Ch. IV, 117–126.

[8] V. Crawford and J. Sobel, "Strategic information transmission," *Econometrica* **50** (1982) 579–594.

[9] J. Farrell, "Meaning and credibility in cheap-talk games," in *Mathematical Models in Economics*, M. Dempster, Ed., Oxford University Press, Oxford, 1988.

[10] F. Forges, "Note on Nash equilibria in repeated games with incomplete information," *International Journal of Game Theory* **13** (1984) 179–187.

[11] F. Forges, "Correlated equilibria in a class of repeated games with incomplete information," *International Journal of Game Theory* **14** (1985) 129–150.

[12] F. Forges, "An approach to communication equilibria," *Econometrica* **54** (1986) 1375–1385.

[13] F. Forges, "Communication equilibria in repeated games with incomplete information," *Mathematics of Operations Research* **13** (1988) 191–231.

[14] R. J. Aumann and S. Hart, Eds., *Handbook of Game Theory with Economic Applications*, North-Holland, Amsterdam, forthcoming.

[15] S. Hart, "Nonzero-sum two-person repeated games with incomplete information," *Mathematics of Operations Research* **10** (1985) 117–153.

[16] *Mathematica*, Reports to the U.S. Arms Control and Disarmament Agency, prepared by *Mathematica*, Inc., Princeton.

[17] J. F. Mertens, "Repeated games," in: *Proceedings of the International Congress of Mathematicians*, Berkeley, California, 1986, pp. 1528–1577.

[18] J. F. Mertens, S. Sorin, and S. Zamir, *Repeated Games*, forthcoming.

[19] R. Myerson, *Game Theory: analysis of conflict*, Harvard University Press, Cambridge, Mass., 1991.

[20] J. Eatwell, M. Milgate, and P. Newman, Eds., *The New Palgrave: a Dictionary of Economics*, McMillan, London, 1987.

[21] S. Sorin, "Some results on the existence of Nash equilibria for non-zero-sum games with incomplete information," *International Journal of Game Theory* **12** (1983) 193–205.

Essay 7

On the Reduced Game of Some Linear Production Games

Daniel Granot[1]

ABSTRACT. In view of the importance of reduced games in cooperative game theory, it is interesting to find out whether they have any meaning in applications. Accordingly we investigate in this paper the Davis–Maschler reduced game of some linear production games. We introduce the extended linear production game model, which generalizes Owen's linear production game model. We show that the Davis–Maschler reduced game of the extended linear production game, calculated at a core point, is a game of the same type, and we provide some sufficient conditions for the nonemptiness of the core of this game. We further investigate the (Davis–Maschler) reduced game of some network games. In particular, we prove that the reduced game of a simple network game, at a core point, is a simple network game and that the core of the reduced game of a simple network game, for any coalition and a core vector, coincides with the set of dual optimal solutions to an associated linear programming problem.

1. Introduction

Reduced games of cooperative games were first defined and investigated by Davis and Maschler [4]. The reduced game property, which was observed therein to be satisfied by the pseudokernel, turned out to be of significant importance in the development of the theory of the kernel, see, e.g., Maschler and Peleg [13], and Maschler, Peleg, and Shapley [14]. A systematic study of the reduced game property was carried out by Aumann and Drèze [1]. Later, Sobolev [20] used the reduced game property to axiomatize the prenucleolus and more recently, Peleg [19] used the reduced game property and the converse reduced game property to axiomatize the prekernel and the core.

---

[1]Faculty of Commerce and Business Administration, The University of British Columbia, Vancouver, B.C. Canada

In view of the importance of reduced games in cooperative game theory, it is interesting whether they have any meaning in applications. In particular, whether a reduced game belongs to the same class of its associated cooperative game. For example, the reduced game of the bankruptcy game (see Aumann and Maschler [2]) is also a bankruptcy game, and Owen [18] has shown that the superadditive cover of the reduced game of an assignment game, computed at a core point, is also an assignment game.

Granot and Maschler [9] investigated the reduced game of some spanning network games. They observed that the reduced game of a minimum cost spanning tree (m.c.s.t.) game is not a m.s.c.t. game. Accordingly, they introduced the class of spanning network games, which extends the classes of m.c.s.t. games (Bird [3], Granot and Huberman [7, 8]), tree games (Megiddo [15]), Steiner tree games (Megiddo [16]) and airport games (Littlechild [12]). They proved that the reduced game of a spanning network game, computed at a core point, is also a spanning network game. They have further shown that the converse reduced game property can be used in some cases to compute the nucleolus of a spanning network game.

In this paper we analyze the reduced game of some linear production games. Apparently, the reduced game of Owen's linear production game [17] is not a linear production game. Accordingly, we introduce a related production game model, to be referred to as the *extended linear production game*, whose reduced game computed at a core point is shown to belong to the same class of games. Further, we provide some sufficient conditions for the nonemptiness of the core of the extended linear production game.

Additionally, we consider the reduced game of some network games (Kalai and Zemel [10, 11]). In particular, it is shown that for each subset S and each core allocation x, the reduced game of a simple network game, calculated at x, is a simple network game whose core coincides with the set of dual solutions to a corresponding linear programming problem.

2. Preliminaries and notation

Let U be a finite set of players. A coalition is a nonempty subset of U. A cooperative game in characteristic function form is the pair $(N; v)$ where $N \subset U$ and $v : 2^N \to R$. We assume that $v(\emptyset) = 0$. A payoff vector for N is a function $x : N \to R$. For each $S \subset N$ we denote $x(S) = \sum(x_j : j \in S)$. A payoff vector for N is efficient in the game $(N; v)$ if $x(N) = v(N)$. It is feasible for $(N; v)$ if $x(N) \leq v(N)$. Let Γ denote the set of all games, and let $\Gamma_0 \subset \Gamma$.

Definition 2.1. A *solution on* Γ_0 is a function, σ, which associates with each game $(N; v) \in \Gamma_0$ a subset, $\sigma(N; v)$, of all feasible payoff vectors for

$(N;v)$. A solution on Γ_0 will be simply referred to as a *solution.*

Definition 2.2. The *core* of a cooperative game $(N;v)$, $C[(N;v)]$, is given by $C[(N;v)] = \{x : x \in R^N,\ x(S) \geq v(S),\ S \subset N,\ x(N) = v(N)\}$.

Definition 2.3. [Davis and Maschler [4].] Let $(N;v)$ be a game, let $S \subseteq N$, $S \neq \emptyset$, and let x be an efficient payoff vector in $(N;v)$. The reduced game with respect to S and x is the cooperative game $(S;v_S^x)$ where

$$v_S^x(T) = \begin{cases} x(S), & T = S \\ 0, & T = \emptyset \\ \max\{v(T \cup Q) - x(Q) : Q \subseteq N \setminus S\}, & \emptyset \neq T \subset S. \end{cases} \tag{2.1}$$

For core vectors x, the reduced game has a simpler representation. Namely,

Lemma 2.4. *Let $(N;v)$ be a cooperative game. Suppose that $C[(N;v)] \neq \emptyset$ and let $x \in C[(N;v)]$. Then, for each $S \subseteq N$ and $T \subseteq S$, $v_S^x(T) = \hat{v}_S^x(T) \equiv$* $\max\{v(T \cup Q) - x(Q) : Q \subseteq N \setminus S\}$.

Proof: In view of Definition 2.3, all we need to show is that $\hat{v}_S^x(\emptyset) = 0$ and $\hat{v}_S^x(S) = x(S)$. Accordingly,

$$\begin{aligned} (i) \qquad 0 &= v(\emptyset) - x(\emptyset) \\ &\leq \max\{v(\emptyset \cup Q) - x(Q) : Q \subseteq N \setminus S\} \\ &= \max\{v(Q) - x(Q) : Q \subseteq N \setminus S\} \\ &\leq 0, \quad \text{since } x \in C[(N;v)]. \end{aligned}$$

Thus, $\hat{v}_S^x(\emptyset) = 0$. Similarly,

$$\begin{aligned} (ii) \qquad x(S) &= v(N) - x(N \setminus S), \quad \text{since } x \text{ is efficient} \\ &= v(S \cup (N \setminus S)) - x(N \setminus S) \\ &\leq \max\{v(S \cup Q) - x(Q) : Q \subseteq N \setminus S\} \\ &\leq x(S), \quad \text{since } x \in C[(N;v)]. \end{aligned}$$

Thus, $\hat{v}_S^x(S) = x(S)$, and the proof is complete. ∎

Definition 2.5. A solution σ on Γ_0 has the reduced game property if whenever $(N;v) \in \Gamma_0$, $S \subseteq N$, $S \neq \emptyset$, and $x \in \sigma(N;v)$, then $(S;v_S^x) \in \Gamma_0$ and $x^S \in \sigma(S;v_S^x)$, where x^S is the restriction of x to S.

The core, among other solution concepts, has the reduced game property, see, e.g., Aumann and Drèze [1] and Peleg [19]. Essentially, the reduced game property is a consistency property. It asserts that if x is a solution which is consistent with the expectations of the players in the game $(N;v)$, then x^S is a solution which should be consistent with the expectations of the players in the reduced game $(S;v_S^x)$.

Next, let us consider the linear production game, introduced by Owen [17]. Therein, each player i, $i \in N$, is in possession of an individual resource vector b^i. The linear production game is the pair $(N;v)$, where for each $\emptyset \neq S \subseteq N$, $v(S) = \max\{cx : Ax \leq \sum(b^i, i \in S),\ x \geq 0\}$. The core of the linear production game is not empty. In fact, as demonstrated by Owen, core points can be computed very efficiently. For example, if u is dual optimal to the linear program associated with N and $y_i = ub^i$, then $y = (y_1, \cdots, y_n)$ is in the core of the linear production game.

3. The extended linear production game

It is easy to verify that the reduced game of a linear production game is not a linear production game. Therefore, in order to have a production game whose corresponding reduced game is a game of the same kind, we introduce below a modified production model having the following parameters:

$N = \{1, \cdots, n\}$ – set of players

A – technology matrix, $A \in R^{m \times p}$

c – revenue vector, $c \in R^p$

b^i – resource vector of player $i, i \in N$, $b^i \in R^m$

t^j, $j \in J$, additional set of resource (technology, knowhow) vectors, $t^j \in R^m$

d_j – cost of purchasing the entire vector $t^j, j \in J$. Purchasing a fraction of t^j is not allowed.

The *extended linear production game* is the pair $(N;v)$, where $v(\emptyset) = 0$ and for every $S \subseteq N$,

$$v(S) = \max\left\{cx - dy : Ax + \sum_{j \in J} y_j t^j \leq \sum_{i \in S} b^i,\ x \geq 0,\ y_j \in \{0,1\}, j \in J\right\}.$$

We will refer to $(N; A; c; b^i, i \in N; t^j, j \in J; d_j, j \in J)$ as the parameters of the extended linear production game.

Proposition 3.1. *The (Davis–Maschler) reduced game of an extended linear production game computed at a core point is an extended linear production game. Specifically, the reduced game on* $S \subseteq N$ *at a core vector* $z \in R^N$ *of an extended linear production game with parameters* $(N; A; c; b^i, i \in N; t^j, j \in J; d_j, j \in J)$ *is an extended linear production game with parameters* $(S; A; c; b^i, i \in S; t^j, j \in J$ *and* $-b^i$, $i \in N \setminus S$; d_j, $j \in J$ *and* z_j, $j \in N \setminus S)$.

Proof: For a core allocation z, the Davis–Maschler reduced game has a simplified representation as shown in Lemma 2.4. The proof then follows from the definition of an extended linear production game. ∎

Thus, a reduced game of an extended linear production game, computed at a core point, is a game of the same kind. However, by contrast with the ordinary linear production game, the core of the extended linear production game could be empty. In the next section we use the reduced game to provide some sufficient conditions for the nonemptiness of the core of an extended linear production game.

4. The core of the extended production game

Clearly, the nonemptiness of the core of the extended linear production game depends on the cost vector, d, associated with the additional set of resources. For example, if the d_j's are sufficiently large then the additional resources will not be used by the various coalitions. In this case, the extended model coincides with the ordinary linear production model in which the additional resources were removed. Therefore, by [17] the core of the extended model in this case is not empty.

In general, we can associate with our extended linear production game $(N; v)$ an *augmented* (ordinary) linear production game $(N \cup J; \bar{v})$ with a corresponding revenue vector c, technological matrix A and a set of players $N \cup J$ such that the resource vector of player i in N (resp., J) is b^i (resp., $-t^i$).

Theorem 4.1. *Let* $z \in C[(N \cup J; \bar{v})]$, *and let* $d_j = z_j$ *for all* $j \in J$ *in the corresponding extended linear production game* $(N; v)$. *Then, the core of the extended linear production game* $(N; v)$ *is not empty. In fact, the restriction of* z, z^N, *to the set of players* N *is in the core of* $(N; v)$.

Proof: By [17], the core, $C[(N \cup J; \bar{v})]$, of the augmented linear production game $(N \cup J; \bar{v})$ is not empty. By Proposition 3.1, the reduced game of $(N \cup J; \bar{v})$ at z is the corresponding extended linear production game $(N;v)$. The proof then follows since the core has the reduced game property. ∎

In view of the discussion preceding Theorem 4.1, the obvious converse to Theorem 4.1 is false. However, a modified converse is valid. Namely,

Theorem 4.2. *Suppose that the core, $C[(N;v)]$, of the extended linear production game is not empty and let $u \in C[(N;v)]$. Then, $y = (u,d) \in R^{|N|+|J|}$ is in the core of the associated augmented linear production game $(N \cup J; \bar{v})$ provided that $\bar{v}(N \cup J) = u(N) + d(J)$.*

Proof: Let $u \in C[(N;v)]$. Then, for all $S \subseteq N$, $u(S) \geq v(S) = \max\{cx - d(R) : Ax + \sum(t^j : j \in R) \leq \sum(b^i : i \in S),\ x \geq 0 \text{ and } R \subseteq J\}$. Thus, for each $Q \subseteq J$ and $S \subseteq N$,

$$
\begin{aligned}
u(S) + d(Q) &\geq \\
&\geq d(Q) + \max\left\{cx - d(R) : Ax \leq \sum_{i \in S} b^i + \sum_{j \in R} -t^j,\ x \geq 0,\ R \subseteq J\right\} \\
&\geq d(Q) + \max\left\{cx - d(Q) : Ax \leq \sum_{i \in S} b^i + \sum_{j \in Q}(-t^j),\quad x \geq 0\right\} \\
&= \max\left\{cx : Ax \leq \sum_{i \in S} b^i + \sum_{j \in Q}(-t^j),\quad x \geq 0\right\} \\
&= \bar{v}(S \cup Q)\,.
\end{aligned}
$$

Thus, if $u(N) + d(J) = \bar{v}(N \cup J)$ then $y = (u,d) \in C[(N \cup J; \bar{v})]$. ∎

As Example 4.3 below demonstrates, for other choices of the cost vector d not taken from the core of the augmented linear production game, the core may be empty.

Example 4.3. Let $N = \{1,2\}$, $c = (2,1)$, $A = \binom{1\ 0}{1\ 1}$, $b^1 = \binom{0}{1}$, $b^2 = \binom{0}{1}$, $J = \{1\}$, $t^1 = -\binom{1}{0}$, and $d_1 = 0$. Then, it is easy to verify that $v(\{1\}) = 2$, $v(\{2\}) = 2$ and $v(\{1,2\}) = 3$, and thus the core of the extended linear production game is empty. Indeed, one can verify that there does not exist a vector z in the core of the associated augmented linear production game in which there is a zero allocation to the player whose resource vector is $(1,0)$.

5. Network games

Network games were introduced in Kalai and Zemel [10, 11] and Dubey and Shapley [5], and further studied, e.g., by Granot and Granot [6]. They arise in network optimization problems in which the players, who control the arcs, seek a fair allocation of the total revenue (or cost) stemming from a flow in the network. More specifically, let $G = (V, A)$ be a directed network with a vertex set V and arc set A. The players, $N = \{1, \ldots, n\}$, control the arcs in G and each arc is controlled by one player. There are two distinct vertices, s and t, designating the source and the sink in the network, respectively. For each $j \in A$, c_j is the (nonnegative) capacity of arc j and r_j designate the revenue (possibly negative) from one unit of flow in arc j. For each $S \subseteq N$ denote by G_S the subnetwork of G consisting only of arcs belonging to S. Then, the pair $(N; v)$, where for each S, $S \subseteq N$, $v(S)$ is the maximum revenue flow from s to t in G_S is a *network game.*

It follows from [17] that if every arc in A belongs to one of the players, i.e., there are no public arcs, then the core of a network game is not empty. On the other hand, one can easily verify that the reduced game of a network game is not a network game. However, since a network game is a special case of a linear production game, one could similarly introduce the class of *extended* network games for which the reduced game of a game in this class, at a core point, is also an extended network game. Similar to linear production games, the core of an extended network game could be empty, and we could derive sufficient conditions for the nonemptiness of the core of an extended network game which would be identical to those derived for extended linear production games.

For the rest of this section we consider the special class of *simple* network games. They are derived from network games by assuming that the capacities of all the arcs in G are one, and were studied by Kalai and Zemel [11], see also Granot and Granot [6].

5.1 SIMPLE NETWORK GAMES

The simple network game $(N; v)$ is derived from the ordinary network game by assuming that the capacity of each arc in $G = (V, A)$ is one, and that each player controls precisely one arc. Thus, $N \subseteq A$. Let $P(A) = A \setminus N$. Then, all arcs in $P(A)$ do not belong to any of the players and will be referred to as *public* arcs.

By definition, for each subset $S \subseteq N$, $v(S)$ is the maximum revenue flow from s to t in the subnetwork, $G_S = (\bar{V}, S \cup P(A))$, of $G = (V, A)$ induced by the set of arcs $S \cup P(A)$. Accordingly, for each $S \subseteq N$,

$$\begin{aligned} v(S) \;=\; \max & \sum_{j \in S \cup P(A)} r_j y_j \\ \text{s.t. } & 0 \leq y_j \leq 1 \quad (j \in S \cup P(A)) \\ & \sum_{j \in \mathrm{IN}_i(G_S)} y_j - \sum_{j \in \mathrm{OUT}_i(G_S)} y_j = 0 \quad (i \in \bar{V} - \{s,t\}) \end{aligned} \tag{5.1}$$

where r_j is the unit revenue of arc j and $\mathrm{IN}_i(G_S)$ (resp., $\mathrm{OUT}_i(G_S)$) is the set of all arcs in G_S which are directed into (resp., out of) node i. We will refer to the last set of constraints in (5.1) as conservation of flow constraints.

We prove below that the reduced game of a simple network game with respect to $S \subseteq N$ and core allocation z is a simple network game, and we provide some sufficient conditions for the nonemptiness of the core of a simple network game (having public arcs). Further, we demonstrate that the core of the reduced game of a simple network game without public arcs, for any subset of players and any core allocation vector, coincides with the set of optimal solutions to an associated linear programming problem.

Proposition 5.1. *Let $(N;v)$ be a simple network game induced by the simple network $G = (V, A)$, $N \subseteq A$, in which the unit revenue of an arc j is $r_j, j \in A$. The reduced game, $(S; v_S^z)$, of $(N;v)$ with respect to $S \subseteq N$ and core allocation z, $z \in C[(N;v)]$, is a simple network game. It is induced by the simple network $G_{S,z} = (V, A)$, in which the unit revenue of an arc $j \in S \cup P(A)$ (resp., $j \in N \setminus S$) is r_j (resp., $r_j - z_j$).*

Proof: It follows from Lemma 2.4 that for each $T \subseteq S$ and core allocation z,

$$\begin{aligned} v_S^z(T) = \max & \sum_{j \in T \cup P(A)} r_j y_j + \sum_{j \in N \setminus S} (r_j - z_j) \\ \text{s.t. } & 0 \leq y_j \leq 1 \qquad (j \in T \cup (A \setminus S) \\ & \sum_{j \in \mathrm{IN}_i(G_{T,z})} y_j - \sum_{j \in \mathrm{OUT}_i(G_{T,z})} y_j = 0 \qquad (i \in V' \setminus \{s,t\}) \\ & y_j \in \{0,1\} \qquad (j \in N \setminus S) \end{aligned} \tag{5.2}$$

where $G_{T,z} = (V', E')$ is the subnetwork of $G_{S,z}$ induced by the set of arcs $E' = T \cup P(A) \cup (N \setminus S) = A \setminus (S \setminus T)$.

It can be easily verified that the constraint matrix in (5.2) is totally unimodular. Therefore, the integrality restrictions therein on y_j, $j \in N \setminus S$, are redundant. Thus, $v_S^z(T)$ is the optimal value of an ordinary maximum revenue flow problem defined over a simple network $G_{T,z}$ in which the unit revenue of an arc $j \in T \cup P(A)$ (resp., $j \in N \setminus S$) is r_j (resp., $r_j - z_j$). ∎

One could easily construct examples in which the core of a simple network game in which there are public arcs is empty. The reduced game of a simple network game is used below to provide some sufficient conditions for the nonemptiness of the core of a simple network game with public arcs.

Proposition 5.2. *Let $(N; v)$ be a simple network game induced by the simple network $G = (V, A)$, in which the set of public arcs $P(A)$, $P(A) = A \setminus N$ is not empty and the unit revenue of arc j is r_j. Let $(A; \tilde{v})$ be a corresponding* augmented simple network game *induced by the same simple network $G = (V, A)$, in which there are no public arcs and the unit revenue of arc j, $j \in N$, is r_j and the unit revenue of arc $j \in P(A)$ is $\tilde{r}_j$. Then, the core of the simple network game $(N; v)$ is not empty if there exists a vector z in the core of the corresponding augmented simple network game $(A; \tilde{v})$ such that $r_j = \tilde{r}_j - z_j$, $j \in P(A)$.*

Proof: By [17], see also [11], the core of a simple network game induced by a network without public arcs is not empty. By Proposition 5.1 the reduced game of $(A; \tilde{v})$, at $N \subset A$ and core allocation z, denoted $(N; v_N^z)$, is also a simple network game. It is induced by the network $G = (V, A)$ with a set of public arcs $P(A) = A \setminus N \neq \emptyset$ and the unit revenue of an arc j in $P(A)$ (resp., $j \in N$) is $\tilde{r}_j - z_j$, (resp., r_j). Thus, if $r_j = \tilde{r}_j - z_j$ for $j \in P(A)$ then $(N; v_N^z)$ coincides with the game $(N; v)$, and the proof follows since the core has the reduced game property. ∎

Similarly, the following result holds for simple network games.

Proposition 5.3. *Suppose the core of the simple network game $(N; v)$, induced by $G = (V, A)$ with revenues $r_j, j \in A$, and public arcs $P(A)$, is not empty, and let $u \in C[(N; v)]$. Then, $y = (u; r_j, j \in P(A)) \in R^{|A|}$ is in the core of the corresponding augmented simple network game $(A; \tilde{v})$, induced by precisely the same network G, provided that $u(N) + \sum_{j \in P(A)} r_j = \tilde{v}(A)$.*

Proof: Similar to the proof of Theorem 4.2 and thus omitted. ∎

For the rest of this section we analyze the core of the reduced game of a simple network game with no public arcs.

5.2 THE CORE OF THE REDUCED GAME OF A SIMPLE NETWORK GAME

Let $(N;v)$ be a simple network game induced by the simple network $G=(V,N)$ with a vertex set V containing the source s and the sink t, arc set N and unit revenue, r_j, for each $j \in N$. The set of players, $N=\{1,\ldots,n\}$, coincides with the set of arcs so that player j owns arc j. We investigate the structure of the core, $C[(S;v_S^z)]$, of the reduced game $(S;v_S^z)$ with respect to a subset S, $S \subseteq N$, and core allocation z, $z \in C[(N;v)]$.

From Proposition 5.1 it follows that the reduced game $(S;v_S^z)$ is induced by the simple network $G_{S,z}=(V,N)$ in which $P(N)=N\setminus S$ are public arcs and the unit revenue of arc j, $j \in S$ (resp., $j \in P(N)$) is r_j (resp., $r_j - z_j$). Further, since there are no public arcs in G, $C[(N;v)] \neq \emptyset$ and thus $C[(S;v_S^z)] \neq \emptyset$ for any $S \subseteq N$ and $z \in C[(N;v)]$.

A path in G is called *simple* if it visits each node in G at most once. Let $P=\{p_1,\cdots,p_k\}$ denote the collection of all simple paths from s to t in G, each considered as a subset of arcs. For each $p \in P$ let

$$r_S(p)=\sum_{j\in p\cap S} r_j + \sum_{j\in p\cap(N\setminus S)} (r_j - z_j)\ .$$

Further, for any $S \subseteq N$ let R^S denote the $|S|$ dimensional space indexed by players in S.

Proposition 5.4. *For each $S \subseteq N$ and $z \in C[(N;v)]$, $C[(S;v_S^z)]$ consists of all vectors $x \in R^S$ satisfying*

$$x(S)=v_S^z(S) \tag{5.3}$$

$$\sum_{j\in p\cap S} x_j \geq r_S(p) \qquad (p \in P) \tag{5.4}$$

$$x_j \geq 0 \qquad (j \in S)\ . \tag{5.5}$$

Proof: (See also [11, Proposition 1]. Clearly, if $x \in C[(S;v_S^z)]$, then x must satisfy (5.3) and (5.5). Further, any such x must satisfy (5.4) for any p for which $p\cap S \neq \emptyset$, while $r_S(p) \leq 0$ for any $p, p\cap S=\emptyset$, since $z \in C[(N;v)]$.

On the other hand, suppose x satisfies (5.3), (5.4), and (5.5). We need to show that for each $T \subset S$, $x(T) \geq v_S^z(T)$. For that purpose, let $y^T=\{y_j : j \in T\cup(N\setminus S)\}$ denote a basic optimal flow solution, yielding an optimal value $v_S^z(T)$, in the subnetwork $G_{T,z}$ of $G_{S,z}$ obtained after deleting the arcs in $S \setminus T$. Since the constraint matrix of a flow problem is totally unimodular, y^T is a $(0,1)$-vector. Thus, y^T induces a set of edge-disjoint simple paths, P', which delivers the optimal flow from s to t. It follows that

$$\begin{aligned} v_S^z(T) &= \sum_{j \in T \cup (N \setminus S)} y_j^T = \sum_{p \in P'} r_S(p) == \sum_{p \in P'} \sum_{j \in p \cap S} x_j \\ &= \sum_{p \in P'} \sum_{j \in p \cap T} x_j \quad \text{(since } p \subseteq T \text{ for each } p \in P'\text{)} \\ &\leq x(T) \ , \end{aligned}$$

where the last inequality holds since the paths in P' are edge-disjoint and x satisfies (5.5). ∎

Lemma 5.5. [Aumann and Drèze [1].] *Let $(N; v)$ be a cooperative game with a nonempty core. Then, for every $z \in C[(N; v)]$ and $S \subseteq N$, $C[(S; v_S^z)] = \{x : x \in R^S$, such that $(x; z_j, j \in N \setminus S) \in C[(N; v)]\}$.*

Proof: For $T \subseteq S$, $Q \subseteq N \setminus S$ and $x \in C[(S; v_S^z)]$,

$$\begin{aligned} x(T) + z(Q) \ &\geq \ v_S^z(T) + z(Q) \\ &= \max\{v(T \cup R) - z(R) : R \subseteq N \setminus S\} + z(Q) \\ &\geq v(T \cup Q) - z(Q) + z(Q) \\ &= v(T \cup Q). \end{aligned}$$

Further, $x(S) + z(N \setminus S) = v_S^z(S) + z(N \setminus S) = z(S) + z(N \setminus S) = z(N)$. Thus, $(x; z_j, j \in N \setminus S) \in C[(N; v)]$. On the other hand, let $(x, x \in R^S; z_j, j \in N \setminus S) \in C[(N; v)]$. Then, since the core has the reduced game property, $x \in C[(S; v_S^z)]$. ∎

For $x \in R^S$, let us denote by $G_{S,z}^x$ the network derived from $G_{S,z}$ by reducing the unit revenue on arcs j, $j \in S$, by x_j. Thus, the unit revenue of arc j in $G_{S,z}^x$ is $r_j - x_j$ (resp., $r_j - z_j$) for $j \in S$ (resp., $j \in N \setminus S$). Proposition 5.6 below demonstrates that it is possible to check very efficiently whether a vector x is contained in the core of the reduced game $(S; v_S^z)$ for any $S \subseteq N$ and $z \in C[(N; v)]$. The proof follows from Lemma 5.5 and [11, Proposition 2]. A direct proof, which is a straightforward extension of the proof of [11, Proposition 2] is included for the sake of completeness.

Proposition 5.6. *Let y be an optimal flow in $G_{S,z}$ and let $x \in R_+^S$ satisfy $x(S) = v_S^z(S)$. Then, $x \in C[(S; v_S^z)]$ if and only if y is optimal for $G_{S,z}^x$ with an optimal value of zero.*

Proof: Since y is optimal for $G_{S,z}$, it is also a feasible flow in $G^x_{S,z}$ with a corresponding total revenue therein equal to

$$\sum_{j\in S} y_j(r_j - x_j) + \sum_{j\notin S} y_j(r_j - z_j) = v^z_S(S) - \sum_{j\in S} y_j x_j. \qquad (5.6)$$

It is easy to show that if there exists an optimal flow, $\bar{y}$, in $G_{S,z}$ for which $\bar{y}_j = 0$ for some $j \in S$, then $x_j = 0$ for every $x \in C[(S; v^z_S)]$. So, let $x \in C[(S; v^z_S)]$. If $x_j > 0$ then $y_j = 1$, and by (5.6), the total revenue corresponding to y in $G^x_{S,z}$ is zero. Therefore, y is optimal if for every simple path p from s to t in $G^x_{S,z}$,

$$\sum_{j\in p\cap S} (r_j - x_j) + \sum_{j\in p\cap(N\setminus S)} (r_j - c_j) \le 0,$$

or, equivalently, if

$$\sum_{j\in p\cap S} x_j \ge \sum_{j\in p\cap S} r_j + \sum_{j\in p\cap(N\setminus S)} (r_j - z_j) = r_S(p). \qquad (5.7)$$

Thus, by Proposition 5.4, y is optimal if $x \in C[(S; v^z_S)]$.

On the other hand, suppose y is optimal with an optimal value zero. Then, by (5.7) $\sum_{j\in p\cap S} x_j \ge r_S(p)$ for any simple path p from s to t. Therefore, since it is assumed that $x \in R^S_+$ and $x(S) = v^z_S(S)$, it follows from Proposition 5.4 that $x \in C[(S; v^z_S)]$. ∎

By [21], a maximum revenue flow problem can be solved in strongly polynomial time. Thus, in view of Proposition 5.6, one could verify in strongly polynomial time whether x is the core of $(S; v^z_S)$ for any $S \subseteq N$ and $z \in C[(N; v)]$.

Let $G = (V, N)$ be a simple network with no public arcs and a corresponding simple network game $(N; v)$. For $z \in C[(N; v)]$ consider the following linear programming problem LP(S),

$$\begin{aligned}
\text{LP}(S):\ \text{Minimize} \ & \sum_{j\in S} x_j \\
\text{s.t.}\ & x_j + \sum_{j\in \text{IN}_i(G)} w_i - \sum_{j\in \text{OUT}_i(G)} w_i \ge r_j \quad (j \in S) \\
& \sum_{j\in \text{IN}_i(G)} w_i - \sum_{j\in \text{OUT}_i(G)} w_i \ge r_j - z_j \quad (j \in N \setminus S) \\
& x_j \ge 0 \quad (j \in S)\,.
\end{aligned}$$

Proposition 5.7. *Let $z \in C[(N; v])$ and $x \in C[(S; v^z_S)]$. Then, there exist values w_i, all nodes i in G, such that $(x, w) = (x_j,\ j \in S;\ w_i,$ all nodes $i)$ is an optimal solution to* LP(S).

Proof: By [11, Theorem 3], the core of a simple network game (without public arcs) consists of the set of optimal vectors x, $x \in R^N$, to the linear programming given below.

$$\text{Minimize} \quad \sum_{j \in N} x_j$$

$$\text{s.t.} \quad x_j + \sum_{j \in \mathrm{IN}_i(G)} w_i - \sum_{j \in \mathrm{OUT}_i(G)} w_i \geq r_j \qquad (j \in N)$$

$$x_j \geq 0 \qquad (j \in N)\,.$$

The proof then follows by Lemma 5.5. ∎

Acknowledgments: This research was partially supported by Natural Science and Engineering Research Council Canada grant A4181 and UBCHSS grant.

6. References

[1] R. J. Aumann and J. H. Drèze, "Cooperative games with coalition structures," *International Journal of Game Theory* **3** (1974) 217–237.

[2] R. J. Aumann and M. Maschler, "Game-theoretic analysis of a bankruptcy problem from the Talmud," *Journal of Economic Theory* **36** (1985) 195–213.

[3] C. Bird, "On cost allocation for a spanning tree: A game theory approach," *Networks* **6** (1976) 335–350.

[4] M. Davis and M. Maschler, "The kernel of a cooperative game," *Naval Research Logistics Quarterly* **12** (1965) 223–259.

[5] P. Dubey and L. S. Shapley, "Totally balanced games arising from controlled programming problems," *Mathematical Programming* **29** (1984) 245–267.

[6] D. Granot and F. Granot, "On some network flow games," *Mathematics of Operations Research* **17** (1992) 792–841.

[7] D. Granot and G. Huberman, "Minimum cost spanning tree games," *Mathematical Programming* **21** (1981) 1–18.

[8] D. Granot and G. Huberman, "On the core and nucleolus of minimum cost spanning tree games," *Mathematical Programming* **29** (1984) 323–348.

[9] D. Granot and M. Maschler, "On some spanning network games," Mimeograph, Faculty of Commerce and Business Administration, U.B.C., 1991.

[10] E. Kalai and E. Zemel, "Totally balanced games and games of flow," *Mathematics of Operations Research* **7** (1982) 476–478.

[11] E. Kalai and E. Zemel, "Generalized network problems yielding totally balanced games," *Operations Research* **30** (1982) 998–1008.

[12] S. C. Littlechild, "A simple expression for the nucleolus in a special case," *International Journal of Game Theory* **3** (1974) 21–29.

[13] M. Maschler and B. Peleg, "The structure of the kernel of a cooperative game," *SIAM Journal of Applied Mathematics* **15** (1967) 569–604.

[14] M. Maschler, B. Peleg, and L. S. Shapley, "The kernel and bargaining set for convex games," *International Journal of Game Theory* **1** (1972) 73–93.

[15] N. Megiddo, "Computational complexity and the game theory approach to cost allocation for a tree," *Mathematics of Operations Research* **3** (1978) 189–196.

[16] N. Megiddo, "Cost allocation for Steiner trees," *Networks* **8** (1978) 1–6.

[17] G. Owen, "On the core of linear production games," *Mathematical Programming* **9** (1975) 358–370.

[18] G. Owen, "The assignment game: The reduced game," Mimeograph.

[19] B. Peleg, "On the reduced game property and its converse," *International Journal of Game Theory* **15** (1986) 187–200.

[20] A. I. Sobolev, "The characterization of optimality principles in cooperative games by functional equations," *Mathematical Methods in the Social Sciences* **6** (1975) 150–165 (in Russian).

[21] É. Tardos, "A strongly polynomial minimum cost circulation algorithm," *Combinatorica* **5** (1985) 247–255.

Essay 8

On Prize Games

Sergiu Hart[1]

ABSTRACT. We consider the class of hyperplane coalitional games (H-games): the feasible set of each coalition is a half-space, with a slope that may vary from one coalition to another. H-games have turned out in various approaches to the value of general non-transferable utility (NTU) games. In this paper we introduce a simple model — *prize games* — that generates the hyperplane games. Next, we provide an axiomatization for the Maschler & Owen [4] consistent value of H-games.

1. Introduction

A *non-transferable utility (NTU) game in coalitional form* is given by specifying the set of outcomes that are achievable by each subset of players ("coalition"). (For formal definitions, see Section 2.) A special class is that of the *transferable utility (TU) games*, where each coalition's feasible set consists of all payoff vectors that add up to no more than a given amount. Geometrically, the feasible sets are half-spaces with normal vector $(1, 1, \ldots, 1)$. Economic models usually lead to NTU-games. TU-games arise if there is a commodity ("money") which is freely transferable among the players and such that one unit of it has exactly the same (marginal) utility to everyone. Single output production models also lead to TU-games.

Another class of NTU-games is that of the *hyperplane (H) games*, where the feasible set of each coalition is a half-space; however, the normal vectors may change from one coalition to another. Thus every TU-game is an H-game and every H-game is an NTU-game; both inclusions are strict.

Hyperplane games have turned up in the study of *value* concepts (see Maschler & Owen [4, 5], Hart & Mas-Colell [3] for the "consistent value"; and Hart & Mas-Colell [2, Theorem C] for the "egalitarian values"). In all

---

[1]Department of Mathematics; Department of Economics; and Center for Rationality and Interactive Decision Theory, The Hebrew University of Jerusalem, Givat-Ram, 91904 Jerusalem, Israel.

Dedicated to Michael Maschler on his 65th birthday.

cases, the main "distinction" of the H-games has been the fact that they are technically more tractable than the general NTU-games. However, no models where H-games arise naturally have been presented.

We will introduce here (in Section 3) a simple model—*prize games*—that yields the class of H-games. Each coalition has one indivisible "prize", which can be given to only one of its members. The feasible set corresponds to all "lotteries" over who gets the prize, and each player is assumed to possess a von Neumann & Morgenstern utility function (over the lotteries). (Such a construction—bargaining over the probabilities of getting a prize—has been used by Roth & Malouf [6] in bargaining experiments.)

The classical value concept is the *Shapley value* [7] for TU-games. It has a "natural" extension[2] to H-games: the *consistent H-value* of Maschler & Owen [4] . Section 4 includes a definition of this value and some of its properties. We then present (in Section 5) a simple axiomatization for the consistent H-value. It generalizes Young's [8] characterization of the Shapley value in the TU-case.

A few notations: $\Re$ is the real line. For a finite set A, the number of elements of A is denoted $|A|$; 2^A is the set of all subsets of A; $\Re^A$ is the $|A|$-dimensional Euclidean space with coordinates indexed by the elements of A; $\Re^A_+$ and $\Re^A_{++}$ are its non-negative and positive orthants, respectively. The origin $(0,0,\ldots,0)$ of $\Re^A$ is written 0^A or just 0. The unit simplex $\{\boldsymbol{x} \in \Re^A_+ : \sum_{i\in A} x^i = 1\}$ of $\Re^A$ is denoted $\Delta(A)$; it is the set of all probability distributions over the set A. For sets, $A \setminus B$ denotes set subtraction, $A \subset B$ is used for *weak* inclusion, and $\emptyset$ is the empty set.

2. Preliminaries

This section includes the basic definitions of games in coalitional form.

A *non-transferable utility (NTU) game in coalitional form* is a pair (N, V), where N is the *set of players* and V is the *coalitional* or *worth function*, associating to every *coalition* $S \subset N$ the non-empty set $V(S) \subset \Re^S$ of all *feasible payoff vectors* for S. An element $x = (x^i)_{i\in S}$ of $V(S)$ is interpreted as follows: there is a feasible outcome for the coalition S whose utility to each player $i \in S$ equals x^i. We will assume that the origin 0^S of $\Re^S$ belongs to $V(S)$ for all S; this is just a convenient normalization.

Two special classes of NTU-games are as follows.

---

[2]It generalizes both Shapley's formula of "expected marginal contribution in a random order," as well as the "consistency property" of Hart & Mas-Colell [1].

- *Hyperplane (H) games*, where each feasible set $V(S)$ is a half-space with positive normal; *i.e.*, for each coalition $S \subset N$, there are $\lambda_S \in \Re^S_{++}$ and $c_S \in \Re$ such that $V(S) = \{x \in \Re^S : \lambda_S \cdot x \leq c_S\}$.
- *Transferable utility (TU) games*, which are H-games with $\lambda_S = (1, 1, \ldots, 1)$ for all S; *i.e.*, there is a real-valued function $v : 2^N \to \Re$ (with $v(\emptyset) = 0$) such that $V(S) = \{x \in \Re^S : \sum_{i \in S} x^i \leq v(S)\}$ for all $S \subset N$. For TU-games, we will write (N, V) or (N, v) interchangeably.

We will call the NTU-game (N, V) *monotonic* if

$$V(S) \times \{0^{T \setminus S}\} \subset V(T) \text{ for all } S \subset T \subset N \,. \tag{2.1}$$

That is, a feasible payoff vector for S completed by 0's to the players in $T \setminus S$ yields a feasible payoff vector for T. For TU-games, this becomes $v(S) \leq v(T)$ for all $S \subset T \subset N$.

3. Prize Games

In this section we will present the model of "prize games", which generate the hyperplane games.

Consider the following setup, which will be called a *(coalitional) prize model.* There is a finite set of players N. Each coalition $S \subset N$ has a "prize" ζ_S. The prize ζ_S is indivisible, and only one member of S can receive it. Moreover, there are no means of utility transfers between the players. Let '0' stand for "no prize", and let $Z := \{\zeta_S : S \subset N\} \cup \{\text{'0'}\}$ be the set of all outcomes. Each player $i \in N$ has a von Neumann & Morgenstern utility function $u^i : \Delta(Z) \to \Re$; recall that $\Delta(Z)$ is the set of probability distributions ("lotteries") over[3] Z. We normalize u^i so that the utility of getting no prize is 0; *i.e.*, $u^i(\text{'0'}) = 0$. For each $i \in S \subset N$, let $z^i_S := u^i(\zeta_S)$ be player i's utility for the prize of S. We assume that $z^i_S \geq 0$ for all i and S : no prize is worse than "no prize". The prize model is actually specified by N and[4] $\boldsymbol{z} := ((z^i_S)_{i \in S})_{S \subset N} \in \prod_{S \subset N} \Re^S_+$; we will denote it $(N, \boldsymbol{z})$.

A prize model $(N, \boldsymbol{z})$ generates a game in coalitional form as follows. The feasible set of each coalition $S \subset N$ consists of all lotteries over which player i in S gets the prize of S; *i.e.*, each $p = (p^i)_{i \in S} \in \Delta(S)$ yields the payoff vector $x = (x^i)_{i \in S}$ with $x^i = p^i z^i_S + (1 - p^i)0 = p^i z^i_S$ for all $i \in S$. Let

[3]Actually, player i's utility needs to be defined only over the prizes of the coalitions containing i; we use the bigger set Z for simplicity of notation.

[4]Such a $\boldsymbol{z}$ is called a *payoff configuration*; it associates a payoff vector to each coalition.

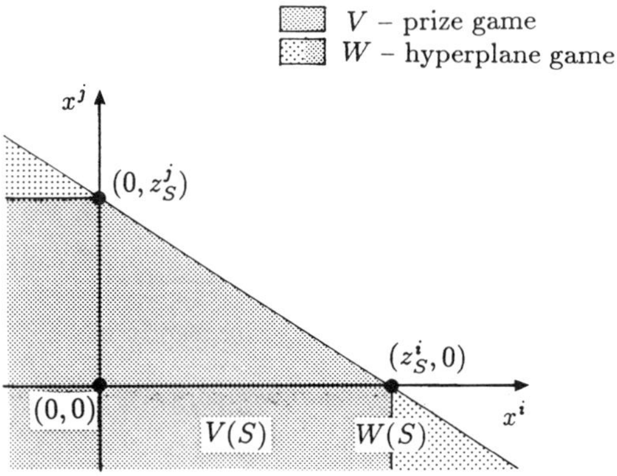

Figure 3.1

$V(S)$ be the comprehensive hull of all these payoff vectors (assume "free disposal"), *i.e.*,

$$V(S) := \{x \in \Re^S : \text{ there is } p \in \Delta(S) \text{ such that } x^i \leq p^i z_S^i \text{ for all } i \in S\}.$$

We will call (N, V) the *coalitional prize game* generated by the prize model $(N, \boldsymbol{z})$.

It is now easy to see that if one considers only non-negative payoffs, then

the class of prize games and the class of hyperplane games coincide

(see Figure 3.1). That is, if (N, V) is a prize game, then there is an H-game (N, W) such that $V(S) \cap \Re_+^S = W(S) \cap \Re_+^S$ for all S (take a hyperplane through the $|S|$ points $\{(0, \ldots, 0, z_S^i, 0, \ldots, 0)\}_{i \in S}$.) Conversely, for every hyperplane game (N, W), there is a prize game (N, V) with $V(S) \cap \Re_+^S = W(S) \cap \Re_+^S$ for all S (define z_S^i by the condition that $(0, \ldots, 0, z_S^i, 0, \ldots, 0)$ lies on the boundary of $W(S)$). Note that if the origin 0^S is an interior point of $V(S)$ for all S or, equivalently, if all the z_S^i are positive rather than just non-negative, then the above correspondence is one-to-one. Moreover, the monotonicity condition (2.1) translates into $z_S^i \leq z_T^i$ for all $i \in S \subset T \subset N$: each player prefers the prize of a larger coalition to that of a smaller one. And finally, a TU-game obtains whenever the utilities of all players coincide; *i.e.*, $z_S^i = z_S^j$ for all $i, j \in S \subset N$.

Thus the coalitional prize setup does indeed provide a simple model for the class of hyperplane games.

4. Value

The *value* of a game is usually interpreted as an a priori evaluation of the outcome of the game. It may be viewed as the utility of each player of participating in the game. (For a recent *non-cooperative* model for the value see Hart & Mas-Colell [3]). The classical value concept is the *Shapley TU-value* [7].

Let (N, v) be a TU-game. For each player $i \in N$ and each coalition S containing i, let $D^i(S, v) := v(S) - v(S \setminus \{i\})$ be player i's *marginal contribution* to the coalition S. Then the Shapley value of player i in the game (N, v) is

$$\mathrm{Sh}^i(N, v) = (1/n!) \sum_{\pi \in \Pi(N)} D^i(P_\pi^i \cup \{i\}, v), \tag{4.1}$$

where $n := |N|$ is the number of players; $\Pi(N)$ is the set of all $n!$ orders[5] on N; and $P_\pi^i := \{j \in N : \pi^{-1}(j) < \pi^{-1}(i)\}$ is the set of players preceding i in the order $\pi \in \Pi(N)$. Equivalently, endow $\Pi(N)$ with the uniform distribution (*i.e.*, each one of the $n!$ orders on N has equal probability $1/n!$), and let $\mathbf{S}$ be the random coalition $P_\pi^i \cup \{i\}$. (Thus, every $S \subset N$ with $S \ni i$ has probability $(s-1)!(n-s)!/n!$, where $s := |S|$.) Then

$$\mathrm{Sh}^i(N, v) = \mathcal{E}[D^i(\mathbf{S}, v)] \tag{4.2}$$

where $\mathcal{E}$ denotes expectation. We denote the vector $(\mathrm{Sh}^i(N, v))_{i \in N} \in \Re^N$ by $\mathrm{Sh}(N, v)$.

The Shapley value has been extended to the class of hyperplane games by Maschler & Owen [4] as follows: Let (N, V) be a hyperplane game and let $\pi \in \Pi(N)$ be an order on N. The *marginal contributions of the players in the order* π, denoted $d_\pi^i \equiv d_\pi^i(N, V)$ (for all $i \in N$), are defined inductively by

$$d_\pi^i := \max\{x^i : (x^i, y_{(i)}) \in V(P_\pi^i \cup \{i\})\}, \tag{4.3}$$

where $y_{(i)} := (d_\pi^j)_{j \in P_\pi^i}$ is the vector of the marginal contributions of all players preceding i in the order π (already defined by induction). That is, d_π^i is the most that player i can get after each preceding player j got his own[6] d_π^j. For $t = 1, 2, \ldots, n$, let $Q_t \equiv Q_t(\pi) := \{\pi(1), \pi(2), \ldots, \pi(t)\}$ be

---

[5]An *order* on N is a one-to-one function $\pi : \{1, 2, \ldots, n\} \to N$; *i.e.*, $(\pi(1), \pi(2), \ldots, \pi(n))$ is a permutation of N.

[6]This suggests that "(marginal) surplus generated by i," or "value added by i" (as in, for instance, "value added tax") are perhaps better names for d^i. We keep the name "marginal contribution" since it is the one that has been classically used for TU-games.

the set of the first t players in the order π. The definition above is plainly equivalent to

$$(d_\pi^k)_{k\in Q_t} \in \partial V(Q_t), \qquad \text{for all } t = 1, 2, \ldots, n, \tag{4.4}$$

where $\partial V(S)$ stands for the boundary of the set $V(S)$ (this is the set of Pareto efficient vectors in $V(S)$). For example, if $N = \{1, 2, \ldots, n\}$ and π is the natural order $(1, 2, \ldots, n)$, then $d_\pi^1 = \max\{x^1 : x^1 \in V(1)\}$; $(d_\pi^1, d_\pi^2) \in \partial V(12)$ determines d_π^2; $(d_\pi^1, d_\pi^2, d_\pi^3) \in \partial V(123)$ determines d_π^3; and so on (see Figure 4.1, for the orders $(1, 2, 3)$ and $(2, 1, 3)$).

The *consistent H-value* φ of Maschler & Owen [4] may be defined by:

$$\varphi^i(N, V) := (1/n!) \sum_{\pi\in\Pi(N)} d_\pi^i(N, V) \tag{4.5}$$

for all $i \in N$: the value to each player is his average (over all orders) marginal contribution. For TU-games $d_\pi^i(N, v) = v(P_\pi^i \cup \{i\}) - v(P_\pi^i) = D^i(P_\pi^i \cup \{i\}, v)$, thus φ is the Shapley value (recall (4.1)). The fact that the boundary of $V(N)$ is a hyperplane guarantees that $\varphi(N, V)$ is an efficient payoff vector, as an average of the efficient vectors $d_\pi = (d_\pi^i)_{i\in N}$. (Again, we write $\varphi(N, V)$ for the vector $(\varphi^i(N, V))_{i\in N} \in \Re^N$.)

Note that in a TU-game, the marginal contribution of i depends only on the set of preceding players P_π^i; in an H-game, it may well depend also on the order of the players inside P_π^i (again, see Figure 4.1). Is there a way to define the marginal contribution of a player i to a coalition S (containing i) in an H-game so that, as in a TU-game, it depends only on the set S? The problem is that one needs to compare two *sets*, $V(S)$ and $V(S \setminus \{i\})$. In general there is no number x^i such that $V(S) = \{x^i\} \times V(S \setminus \{i\})$ (unless—as in a TU-game—the two sets are "parallel", meaning that $(\lambda_S^j)_{j\in S\setminus\{i\}}$ and $\lambda_{S\setminus\{i\}}$ are proportional).

We are thus led to the idea to "summarize" the possibilities of $S \setminus \{i\}$ by one payoff vector, rather than having a whole set. The natural candidate is, of course, the value of the subgame[7] $(S \setminus \{i\}, V)$. We then define the marginal contribution of player i as the most he can get in $V(S)$ when the players in $S \setminus \{i\}$ get the value of $(S \setminus \{i\}, V)$.

Formally, let (N, V) be a hyperplane game, and assume by induction that the values $\varphi(T, V) \in \Re^T$ of its subgames (T, V), for all $T \subsetneqq N$, are given. Let $i \in N$ be a player and let S be a coalition containing i. We define

[7] Given a game (N, V) and a coalition $T \subset N$, we write (T, V) for the *subgame* whose player set is T and whose coalitional function is the restriction of V to the subsets of T (*i.e.*, to 2^T).

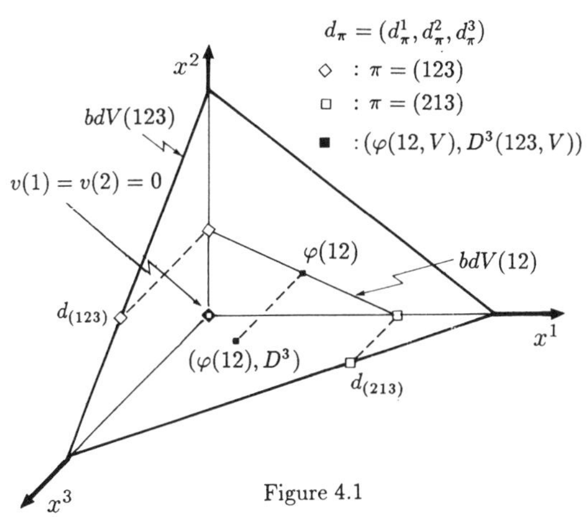

Figure 4.1

the *marginal contribution of player i to the coalition S* as[8]

$$D^i(S, V) := \max\{x^i : (x^i, \varphi(S \setminus \{i\}, V)) \in V(S)\}. \tag{4.6}$$

Equivalently, $D^i(S, V)$ is uniquely determined by the condition

$$(D^i(S, V), \varphi(S \setminus \{i\}, V)) \in \partial V(S). \tag{4.7}$$

Note that in order to define the marginal contributions one needs to know the values of the subgames (T, V) for all subsets T. For TU-games we get of course $D^i(S, v) = v(S) - v(S \setminus \{i\})$, since $\varphi = \mathrm{Sh}$ yields efficient payoff vectors (moreover, in this case of TU-games we don't need to know $\varphi(T, v)$, only that it is efficient).

Proposition 4.1. *The consistent H-value φ satisfies*

$$\varphi(N, V) = (1/|N|) \sum_{i \in N} (D^i(N, V), \varphi(N \setminus \{i\}, V)); \tag{4.8}$$

and

$$\varphi^i(N, V) = \mathcal{E}[D^i(\mathbf{S}, V)] \tag{4.9}$$

[8]Note that d_π^i was called "marginal contribution of i *in the order* π," whereas $D^i(S)$ is the "marginal contribution of i *to the coalition* S." In H-games, these are distinct notions (again, see Figure 4.1).

for every hyperplane game (N, V) *and* $i \in N$. *Moreover, if* (N, V) *is a monotonic hyperplane game, then* $\varphi(N, V)$ *is non-negative and it depends only on the non-negative parts of the feasible sets* (i.e., *if* $V(S) \cap \Re^S_+ = W(S) \cap \Re^S_+$ *for all* S *then* $\varphi(N, V) = \varphi(N, W)$).

> *Proof:* [9] By induction: In (4.5), consider all orders where player i comes last. The (conditional) average marginal contribution of each $j \neq i$ is $\varphi^j(N \setminus \{i\}, V)$ by the induction hypothesis; efficiency then implies that the i-th coordinate is the "remainder," namely, $D^i(N, V)$ (see (4.7)). This proves (4.8), from which (4.9) follows, again by induction. The "moreover" statements for monotonic games are immediate by any of the formulas (4.5), (4.8) or (4.9). ∎

The interpretation of formula (4.9) (which is the same as (4.2) for the Shapley value) is clear: The value to each player is his expected marginal contribution. As for formula (4.8), it may be viewed as follows: One takes the value payoff vectors of the n subgames with $n-1$ players, "completes" each one to an efficient payoff vector for N (see (4.7)), and then averages these vectors – to obtain the value of the n-player game. Thus the value to player i is the average of his marginal contribution $D^i(N, V)$ to the grand coalition N and of his values when each one of the other players "drops out" (*i.e.*, $\varphi^i(N \setminus \{j\}, V)$ for all $j \neq i$). Note that, in particular, we obtain that the Shapley TU-value satisfies (4.8).

It is interesting to interpret the various formulas for the consistent H-value in the context of the prize games model. The value associates to a prize game (N, z) a probability distribution $\rho(N, z) \in \Delta(N)$, with $\rho^i(N, z)$ being the probability that i gets the "grand" prize ζ_N. The connection between ρ and φ is of course $\varphi^i(N, V) = \rho^i(N, z) z^i_N$ for all $i \in N$ (where (N, V) is the corresponding coalitional prize game). We will refer to ρ as the *value probability*.

Consider for instance formula (4.8), and assume that the prize game is monotonic. As noted in Section 3, this implies in particular that $z^j_N \geq z^j_{N\setminus\{i\}}$ for all $j \neq i$. Let $\theta^j_i := z^j_{N\setminus\{i\}}/z^j_N$; *i.e.*, player j is indifferent between getting the prize $\zeta_{N\setminus\{i\}}$ of $N \setminus \{i\}$ for sure, and getting the grand prize ζ_N of N with probability θ^j_i (and nothing with probability $1 - \theta^j_i$).

Write $\rho^j_i := \rho^j(N \setminus \{i\}, z)$ for the value probability of j in the $N \setminus \{i\}$ game; then j is indifferent between getting $\zeta_{N\setminus\{i\}}$ with probability ρ^j_i, and getting the grand prize ζ_N with probability $\rho^j_i \theta^j_i$. The remaining probability, namely $\delta^i := 1 - \sum_{j \in N\setminus\{i\}} \rho^j_i \theta^j_i$, is i's *marginal probability* (indeed,

---

[9] Note that formula (4.8) is essentially formula (5.1) in Maschler & Owen [4]. In (4.9), $\mathbf{S}$ is the random coalition of i together with all players preceding i in a random order, and $\mathcal{E}$ denotes expectation (see (4.2) above).

$\delta^i z^i_N = D^i$; see (4.7)). Then $\rho^i = (1/n)(\delta^i + \sum_{j \in N \setminus \{i\}} \rho^i_j \theta^i_j)$; *i.e.*, the value probability of i in $(N, \boldsymbol{z})$ is the average of his marginal probability δ^i and of his value probabilities $\rho^i_j \theta^i_j$ in the subgames $(N \setminus \{j\}, \boldsymbol{z})$, evaluated in terms of the grand prize. Similar interpretations may be given to the other formulas, namely (4.5) and (4.9).

5. Axiomatization

There are to date four approaches to the consistent H-value: First, generalizing the random order formulas for the Shapley value (see Maschler & Owen [4], and also the previous Section). Second, using an appropriate "consistency" or "reduced game" property (Maschler & Owen [4]). Third, a dynamic procedure (Maschler & Owen [4]). And fourth, via a noncooperative foundation (Hart & Mas-Colell [3]). We will present here another approach; it is axiomatic, and based on the "marginality" idea introduced by Young [8] for TU-games.

A *solution function* ψ on a class Γ of games is a mapping that associates to every game (N, V) in Γ a payoff vector[10] $\psi(N, V)$ that is feasible for the grand coalition N; *i.e.*, $\psi(N, V) \in V(N)$. We only consider H-games; the domain Γ will thus always be included in the class of all H-games.

The solution function ψ is *efficient* if it always yields (Pareto) efficient payoff vectors; *i.e.*, $\psi(N, V) \in \partial V(N)$ for every game $(N, V) \in \Gamma$. Two players $i, j \in N$ are called *substitutes* in a game (N, V) if for all $S \subset N \setminus \{i, j\}$, all $x_S \in \Re^S$ and all $\xi \in \Re$, we have $(x_S, \xi) \in V(S \cup \{i\})$ if and only if $(x_S, \xi) \in V(S \cup \{j\})$. The solution function ψ is *symmetric*[11] if $\psi^i(N, V) = \psi^j(N, V)$ whenever i and j are substitutes in (N, V).

Efficiency and symmetry are standard requirements. We now introduce another postulate. The solution function ψ satisfies *marginality* if $\psi^i(N, V) = \psi^i(N, W)$ whenever the marginal contributions of player $i \in N$ are identical in the two games (N, V) and (N, W); *i.e.*, if $D^i(S, V) = D^i(S, W)$ for all $S \subset N$ with $S \ni i$, then $\psi^i(N, V) = \psi^i(N, W)$. Marginal contributions to a coalition are defined as in the previous Section (see (4.7)) using the solution ψ for the subcoalitions, namely

$$D^i(S, V) := \max\{x^i : (x^i, \psi(S \setminus \{i\}, V)) \in V(S)\} \; ; \tag{5.1}$$

equivalently, $D^i(S, V)$ is uniquely determined by the condition

$$(D^i(S, V), \psi(S \setminus \{i\}, V)) \in \partial V(S). \tag{5.2}$$

---

[10]We are thus considering only *one-point* solutions.

[11]This is also called "equal treatment property."

Marginality says that if a player has the same marginal contributions in two games, then his value must be the same in the two games.

We can now state the main characterization result.

Theorem 5.1. *Let ψ be a solution function on the class of hyperplane games. Then ψ satisfies efficiency, symmetry and marginality if and only if ψ is the consistent H-value.*

Marginality implies that the value is a function of the marginal contributions only. However, it could be any function, not necessary linear. It is remarkable that efficiency and symmetry suffice to determine this function uniquely (as the average (4.9)).

Proof: The three axioms applied to TU-games uniquely characterize the Shapley value there; this is the result of Young [8, Theorem 2]. For general hyperplane games, we will prove by induction that ψ must coincide with the consistent H-value φ. Let (N, V) be a hyperplane game, and assume that $\psi(S, V) = \varphi(S, V)$ for all $S \subsetneqq N$. Fix $i \in N$. Define a TU-game (N, w) as follows: $w(S) := 0$ if $i \notin S$ and $w(S) := D^i(S, V)$ if $i \in S$. Then $D^i(S, w) = w(S) - w(S \backslash \{i\}) = D^i(S, V)$ for all S containing i, which by marginality implies that $\psi^i(N, w) = \psi^i(N, V)$. But (N, w) is a TU-game, therefore (by (4.2) and then (4.9)) $\psi^i(N, V) = \psi^i(N, w) = \mathrm{Sh}^i(N, w) = \mathcal{E}[D^i(\mathbf{S}, w)] = \mathcal{E}[D^i(\mathbf{S}, V)] = \varphi^i(N, V)$. The converse, *i.e.*, that φ satisfies the three axioms, is immediate (again, see (4.9)). ∎

The same result holds for *monotonic* games.

Theorem 5.2. *Let ψ be a solution function on the class of monotonic hyperplane games. Then ψ satisfies efficiency, symmetry and marginality if and only if ψ is the consistent H-value.*

Proof: First, note that the characterization of Young [8] for the Shapley TU-value holds when restricted to the space of monotonic TU-games (the proof described there for superadditive games actually applies to any cone of TU-games that includes all the unanimity games). Next, the only change needed in the proof of Theorem 5.1 is to make w monotonic. This is done as follows: for each S not containing i define $w(S) := \sum_{T \subsetneqq S} D^i(T \cup \{i\}, V)$ and $w(S \cup \{i\}) := \sum_{T \subset S} D^i(T \cup \{i\}, V)$. Since (N, V) is monotonic we have $D^i(T \cup \{i\}, V) \geq 0$, from which the monotonicity of (N, w) follows immediately. ∎

Remarks 5.3.

(i) Efficiency and symmetry are actually needed only for TU-games.

(ii) The results here may be extended to general (non-hyperplane) NTU-games; with an appropriate definition of marginal contributions, the three axioms of efficiency, symmetry and marginality characterize the consistent NTU-value of Maschler & Owen [5].

Acknowledgments: Research partially supported by the National Science Foundation. I wish to acknowledge useful discussions on these topics with Robert J. Aumann, Michael Maschler, Andreu Mas-Colell and Eyal Winter.

6. References

[1] S. Hart and A. Mas-Colell, "Potential, value and consistency," *Econometrica* **57** (1989) 589–614.

[2] S. Hart and A. Mas-Colell, "Egalitarian solutions of large games: I. A continuum of players," DP #1, The Center for Rationality and Interactive Decision Theory, The Hebrew University of Jerusalem, mimeo, 1992.

[3] S. Hart and A. Mas-Colell,"A model of n-person noncooperative bargaining," DP #10, The Center for Rationality and Interactive Decision Theory, The Hebrew University of Jerusalem, mimeo, 1992.

[4] M. Maschler and G. Owen, "The consistent Shapley value for hyperplane games," *International Journal of Game Theory* **18** (1989) 389–407.

[5] M. Maschler and G. Owen, "The consistent Shapley value for games without side payments," in: *Rational Interaction*, R. Selten, Ed., Springer-Verlag, 1992, pp. 5–12.

[6] A. E. Roth and M. K. Malouf, "Game-theoretic models and the role of information in bargaining", *Psychological Review* **86** (1979) 574–594.

[7] L. S. Shapley, "A Value for n-Person Games," in: *Contributions to the Theory of Games II, Annals of Mathematics Studies* **28**, H. W. Kuhn & A. W. Tucker, Eds., Princeton University Press, 1953, pp. 307–317.

[8] H. P. Young, "Monotonic solutions of cooperative games," *International Journal of Game Theory* **14** (1985) 65–72.

Essay 9

On Probabilistic Machines, Bounded Rationality and Average-Case Complexity

Nimrod Megiddo[1]

ABSTRACT. The analogs of pure, mixed and behavior strategies in the context of algorithms are studied. It is shown that probabilistic machines are more powerful than probability distributions over deterministic ones, that best response may sometimes requires randomization, and that if nature's choices are computable then there exists a deterministic best response.

1. Introduction

In traditional game theory, deterministic strategies are called pure, and probability distributions over pure strategies are called mixed strategies. A strategy which makes probabilistic choices at every decision point is called a behavior strategy. It is well known that a mixed strategy is at least as powerful as a behavior strategy and the two are of equivalent power if the game has perfect recall [1]. Also, a best response to the opponent's mixed strategy can be played without randomization.

In recent years, there has been a growing literature on playing games through computing machines (see, for example, [2]), attempting to understand bounded rationality. On the other hand, complexity theory sometimes uses ideas from game theory (see, for example, [3]).

With some reservations, algorithms can be viewed as strategies, so concepts similar to mixed and behavior strategies can be defined with respect to algorithms. We show here that algorithms exhibit phenomena which are quite different from the ones we know from game theory. As usual in the theory of complexity, an algorithm is a procedure for solving a family of problem instances rather than a single one.

---

[1]IBM Almaden Research Center, 650 Harry Road, San Jose, California 95120-6099, and School of Mathematical Sciences, Tel Aviv University, Tel Aviv, Israel.

2. Models and results

Let $\mathcal{M}$ denote the set of all *deterministic* Turing machines M such that, given any natural number n as input, the machine M computes an n-bit number $M(n)$. Let $\mathcal{P}$ denote the analogous set of *probabilistic* Turing machines.[2]

THE BASIC GAME

The definition of the basic game considered here is inspired by the concept of worst-case complexity of a computational problem. Player I (the player) designs a deterministic algorithm for the problem, and player II (the adversary) picks inputs for this algorithm for each input size. Thus, the player chooses a deterministic machine $M \in \mathcal{M}$, and the adversary chooses a function A which assigns to every natural n, an n-bit number $A(n)$. The adversary then pays the player an amount $C_n = C_n(M, A) = 2^n$, if $M(n) = A(n)$, and $C_n(M, A) = 0$ otherwise.

It is trivial to see that the player cannot secure for himself any positive payoff for any input size, since the adversary may choose $A(n) \neq M(n)$ for every n.

MIXED STRATEGIES

A mixed strategy is essentially a probability distribution over the deterministic strategies. If mixed strategies are allowed, then this game is played as follows. The player chooses any probability distribution $\pi : \mathcal{M} \to R$, i.e., $\pi(M) \geq 0$ and $\sum_{M \in \mathcal{M}} \pi(M) = 1$. Suppose the adversary now chooses his function $A(n)$, knowing this probability distribution. The expected payoff is evaluated with respect to π, i.e.,

$$E_n = E_n(\pi, A) = \mathcal{E}\, C_n(M, A) = \sum_{M \in \mathcal{M}} \pi(M) C_n(M, A) \ .$$

We note that since the adversary knows π, he cannot benefit from randomizing his choice of the mapping A, i.e., for every n the adversary has an optimal choice of a number $A(n)$, namely, one with minimum probability.

Ideally, the player would like to choose a probability distribution over $\mathcal{M}$ which would induce, for every n, a uniform distribution over the n-bit numbers. Given such a probability distribution, the adversary would be

[2] A probabilistic Turing machine is one that can also flip a coin and read its outcome.

indifferent, and the value of the game would be 1 for every n. However, such a probability distribution does not exist! The proof of the latter claim is as follows. Let π be any probability distribution over $\mathcal{M}$. Since $\mathcal{M}$ is countable, there exists an $M \in \mathcal{M}$ such that $\epsilon \equiv \pi(M) > 0$. It follows that for every n such that $2^{-n} < \epsilon$, the probability that the number $M(n)$ will be chosen is greater than 2^{-n}, hence there exists another n-bit number k_n which is chosen with probability less than 2^{-n}. Since the adversary knows the distribution π, he can find such a number k_n and choose $A(n) = k_n$. Thus, for every such n, the value of the game is less than 1.

In fact, we can prove a much stronger result. Let δ be any positive number. For any probability distribution π over $\mathcal{M}$, there exists a finite number K and machines $M_1, \ldots, M_K \in \mathcal{M}$ such that

$$\sum_{i=1}^{K} \pi(M_i) > 1 - \delta \ .$$

Let n be such that $2^n > K$. It follows that there exists an n-bit number k_n whose probability is less than $\delta/(2^n - K)$. If the adversary chooses $A(n) = k_n$ then the limsup of the expected payoff, as n tends to infinity is no more than δ. Since δ can be any positive number, the limit is indeed zero!

Note that the actual sequence of choices of numbers computed by the sampled deterministic machine is of course computable. The adversary can choose a noncomputable sequence $A(n)$. On the other hand, we have not assumed that the probability distribution itself is computable. If we assume the probability distribution is computable, and the adversary is given a program that computes $\pi(M)$ for every M, then we can also restrict the adversary to choose computable sequences $A(n)$; the expected payoff still tends to zero since in this case there exists a program which identifies an appropriate k_n for every n.

It is interesting to consider the rate of convergence of the value to zero. We now give an example. Given a natural number n, let $m = 2^n$. For simplicity (and with no loss of generality), let's change notation so that the object of the game is: given m, produce a number from the set $S_m = \{1, 2, \ldots, m\}$. We first construct a sequence of machines $M_1, M_2, \ldots$ as follows. The machine M_j computes a number $k \in S_m$ such that $j \equiv k \bmod m$. Let M_j be chosen with probability

$$p_j = \frac{j^{-c}}{\zeta(c)} \ ,$$

where $c > 1$ is a constant and $\zeta(c) = \sum_{i=1}^{\infty} i^{-c}$. It follows that, given m, a number $k \in S_m$ is chosen with probability

$$P_m(k) = \sum_{i=0}^{\infty} p_{k+im} \ .$$

The one with the smallest probability in S_m is m itself:

$$P_m(m) = \sum_{i=0}^{\infty} p_{m(i+1)} = m^{-c} .$$

The payoff is therefore, m^{1-c}, which tends to zero as m tends to infinity. By choosing c sufficiently small we can guarantee convergence at a polynomial rate with any degree. Of course, slower convergence can be achieved by choosing the $\pi(M_i)$ converging more slowly than any series of the form i^{-c}.

Behavior strategies

A behavior strategy is essentially a strategy that allows probabilistic decisions at any decision point. This concept corresponds to the notion of a probabilistic Turing machine in our game. Thus, if behavior strategies are allowed, then the game is played as follows. The player chooses a probabilistic machine $\boldsymbol{M} \in \mathcal{P}$ and then the adversary chooses a function A as above. The payoff is now a random variable $\boldsymbol{C}_n = \boldsymbol{C}_n(\boldsymbol{M}, A)$ whose value depends on the coin tosses of $\boldsymbol{M}$, so its expected payoff is evaluated with respect to this distribution.[3]

It is obvious that we have an optimal probabilistic Turing machine, namely, one which for every n samples an n-bit number simply by writing down n random bits. The value of the game is 1 for every n.

Distributions of inputs

We also consider games inspired by the concept of average-case complexity. First, suppose nature moves first and selects for every n, any probability distribution over the n-bit numbers. The player is then informed of these selections and then chooses either $M \in \mathcal{M}$ or $\boldsymbol{M} \in \mathcal{P}$, depending on the version being played. Next, for every n a number $A(n)$ is sampled from the probability distribution selected by nature, and the chosen machine produces a number, so payoffs can be determined as above.

If the player is allowed to choose a probabilistic machine, then the one discussed above guarantees the value of 1 for every n, regardless of nature's choice of distribution. However, if the player has to choose a deterministic machine then the value may be less than 1. For example, even if nature's choices are deterministic, but the sequence $A(n)$ is not computable, then the player will get 0 for infinitely many values of n.

---

[3]Mixtures of behavior strategies can also be defined.

Computable distributions of inputs

We now claim that if nature's choice of probability distribution is computable, then the player has an advantage. The variant we consider is as follows. Nature moves first and selects a Turing machine[4] which computes, for every n, a probability distribution over the n-bit numbers. The player is given the full description of the machine selected by nature and the game proceeds as above.

In this special case the player can design a deterministic machine M that analyzes nature's choices for any value of n and computes the best response to that probability distribution, namely, a number $M(n)$ of maximum probability. This yields an expected payoff of at least 1 for every n. Of course, nature can force the expected payoff to be not greater than 1.

3. Summary

To summarize, we have established the following facts, which are quite different from what traditional game theory suggests:

1 Probabilistic machines are more powerful than probability distributions over deterministic ones.

2 Randomization may be necessary in order to respond optimally to a known probability distribution chosen by nature.

3 If nature's choice of probability distribution is computable, then there exists a deterministic best response.

Acknowledgments: Thanks to Nathan Linial and other attendees of the IMSSS Summer Workshop on Bounded Rationality, for stimulating discussions on the material in the paper.

4. References

[1] H. W. Kuhn, "Extensive games and the problem of information," in: *Contributions to the Theory of Games I*, H. W. Kuhn and A. W. Tucker, eds., Annals of Mathematics Studies No. 28, Princeton University Press, Princeton, New Jersey, 1950.

[4] Here it does not matter whether this machine is deterministic or probabilistic.

[2] N. Megiddo and A. Wigderson, "On play by means of computing machines," in: *Theoretical Aspects of Reasoning about Knowledge*, J. Y. Halpern, ed., Morgan Kaufmann Publishers, Inc., Los Altos, California, 1986, pp. 259–274.

[3] A. C. Yao, "Probabilistic computations: toward a unified measure of complexity," in: Proceedings of the 18th Annual IEEE Symposium on Foundations of Computer Science (1977), IEEE Computer Society Press, Long Beach, CA, 1977, pp. 222–227.

Essay 10

Consistent Demands for Coalition Formation

Benny Moldovanu[1] and Eyal Winter[2]

ABSTRACT. In equilibria of a bargaining model of coalition formation and payoff division (see Selten, 1981) players appear to form demands for their participation in a coalition. These demands have several appealing features. We characterize the sets of semi-stable and stable demand vectors for general NTU games using consistency requirements that relate the demands formed in the initial game to demands formed by subgroups considering their outside options.

1. Introduction

Selten [13] analyses a non-cooperative sequential model of bargaining between n players that face a situation described by an underlying game in coalitional form with side payments (or TU game). A single coalition (among the many potential ones) can form, and its members must agree on a division of the available payoff to that coalition. Examples are the formation of a ruling coalition in a parliament, or the exploitation of a natural resource by one of many possible consortia.

An interesting phenomenon arises in Selten's model: In equilibrium the players appear to set demands (or reservation prices) for their participation in any coalition. If a coalition indeed forms then it divides the available payoff exactly according to these demands. It is also shown that the demand vectors—that express the players' expectations were they to be members in a forming coalition—have several properties: 1. The players form the demands in such way that there are no "leftovers" in any coalition. 2. Each player sets his demand such that he can exhibit at least one coalition, including himself, that can satisfy the demands of its members. 3. With some additional refinement concepts, the demands are such that no player

[1]Department of Economics, University of Michigan, Ann Arbor, MI 48108
[2]Department of Economics, The Hebrew University, Jerusalem, Israel.

i is dependent on player j in the sense that i always needs j to form a coalition satisfying its members' demands, but j does not need i in order to do so.

The presence of a game with side-payments implies quite restrictive assumptions about the utility functions of the players. Selten's model can be easily adopted for the situation where the players face a game without side payments (or NTU game) and all the results continue to hold.

Independently from the non-cooperative implementation, demand vectors exhibiting the above mentioned properties have been defined and studied by Albers [1] and Bennett [3]. Bennett and Zame [5] prove existence of such vectors for a large class of NTU games. In Moldovanu and Winter [11] we study the relationships between demand vectors and the core of a convex game, and we use these relationships for a new characterization of convex games.

The purpose of this paper is to study consistency properties of the demand vectors. The consistency conditions relate here to the possibility of partial implementation of the outcome by subgroups of players.

For example, imagine a parliament where parties A, B, and C contemplate the formation of a ruling coalition and express demands concerning the common policy, portfolios, etc. Assume that it is agreed that C will obtain its demand. Then consistency requires that the original demands of A and B should remain reasonable if they contemplate cooperating (together or separately) with other remaining parties, given the expressed demands of these parties. Higher demands of A and B can be treated as incredible threats. Conversely, if A and B form demands based on the expectations they have in other potential ruling coalitions then these demands constitute a reasonable basis for the present negotiation.

Consistency requirements have played a major role in the analysis and characterization of solution concepts in cooperative game theory. Due to the technical difficulties connected to the definition and existence of solution concepts there have been relatively fewer application of consistency ideas in the framework of coalitional games with non-transferable utility. We refer the reader to Peleg [12] (the core), Hart and Mas-Colell [7] (the egalitarian solution), Moldovanu [10] (stable equilibria for bargaining in two-sided markets), Thomson and Lensberg [14] (Nash and other bargaining solutions) and Maschler and Owen [9] (value for hyperplane games). We note that demand vectors for TU games are characterized in Winter [15].

Michael Maschler's work has greatly influenced the research on the topics discussed in this paper. Firstly, the intuition captured by the bargaining set (see Aumann and Maschler, 1964) has been crucial in the development of one of the demand concepts analyzed here (see below — the stable demand correspondence). For a detailed discussion of this relation see Bennett

[4]. Secondly, reduced games for coalitional (TU) games were first defined and analyzed in Davis and Maschler [6]. The concept of reduced games constitutes the foundation on which all subsequent work on consistency properties and axiomatic characterizations of solution concepts is based.

This paper is organized as follows: At the end of this section we introduce some notation. In Section 1 we define the sets of semi-stable and stable demand vectors for NTU games. In Section 2 we study consistency properties of these sets. In Section 3 we characterize the sets of semi-stable and stable demand vectors using the consistency requirements. In Section 4 we briefly compare our results to other characterizations of solution concepts for NTU games.

Notation: Let N be a finite set of players. A coalition is a non-empty subset of N. A payoff vector for N is a function $x : N \to \Re$, thus $\Re^N$ is the set of all payoff vectors. $\Re^S_+$ is the restriction of $\Re^S$ to vectors with non-negative coordinates. x^S denotes the restriction of x to members of S. $x(S)$ denotes the sum $\sum_{i \in S} x^i$. O^S denotes the vector in $\Re^S$ with all coordinates equal to zero.

Let $x, y \in \Re^S$. We write: $x \geq y$ if $x^i \geq y^i$ for all $i \in S$; $x > y$ if $x \geq y$ and $x \neq y$; $x \gg y$ if $x^i > y^i$ for all $i \in S$.

Let $A \subseteq \Re^S_+$. The set A is *comprehensive* if $x \in A$, $y \in \Re^S_+$ and $x \geq y$ imply together that $y \in A$. The interior of A relative to R^S_+ is denoted by $\operatorname{int} A$. The set $A \setminus \operatorname{int} A$ is denoted by ∂A. The set A is *non-level* if

$$(x, y \in \partial A) \wedge (x \geq y) \Rightarrow x = y \; . \tag{1.1}$$

2. Semi-stable and Stable Demand Vectors for NTU Games

Definition 2.1. An *NTU game* is a pair (N, V) where N is a non-empty set of players and V is a function that assigns to each coalition S in N a subset $V(S)$ of $\Re^S_+$, with the following properties:

(i) $V(S)$ is non-empty and compact.

(ii) $V(S)$ is comprehensive relative to $\Re^S_+$.

(iii) $V(S)$ is non-level.

Definition 2.2. Let (N, V) be an NTU game and let $x \in \Re^N$. Denote by $F_i(x, V)$ the following set:

$$F_i(x, V) = \{S \mid S \subseteq N; i \in S \text{ and } x^S \in V(S)\} \tag{2.1}$$

x is called a *semi-stable demand vector* for (N, V) if it satisfies:

(i) $x^S \notin \operatorname{int} V(S)$ for every coalition S in N. (*Maximality*).

(ii) $F_i(x, V) \neq \emptyset$ for every player $i \in N$. (*Feasibility*).

x is called a *stable demand vector* for (N, V) if it is semi-stable and additionally it satisfies

(iii) For every $i, j \in N$, $i \neq j$, it is not the case that $F_i(x, V) \subset F_j(x, V)$. (*Independence*).

The terminology is due to Selten [13]. He refers there to games with side payments. It is interesting to note the set of stable demand vectors carries over the intuition of the bargaining set M_i^1 (see Aumann and Maschler [2]) from the set of imputations to the set of semi-stable demand vectors. This relation is discussed in Bennett [4]. Note that in that article, and in some other papers as well, the set of semi-stable demand vectors is called "the set of aspirations," and the set of stable demand vectors is called "the aspiration bargaining set."

Interpretation: Each player sets a demand for his participation in a coalition. The maximality condition (i) requires that the demands are maximal in the sense that no one can raise his demands without excluding himself from any coalition that can satisfy the demand of its members. The feasibility condition (ii) ensures that each player can propose at least one coalition that includes him and that satisfies the demand of its members. The independence condition (iii) ensures that no player is dependent on another one in the sense that the first player needs always the second to ensure feasibility of his demand, but the second player can find a coalition that satisfies its member's demands, but excludes the first player.

The demands can be thought as reservation prices set by the players for their participation in any coalition. These demand concepts *are not* solution concepts in the usual sense of the word. It may be the case that, for a particular demand vector, there exists no coalition structure such that the demand vector is feasible for that structure. Nevertheless, if a coalition actually forms, the division of the payoff according to the respective demands may constitute a reasonable outcome. By abusing a little the terminology we will refer to the demand concepts as "solution concepts"—mainly to ease the comparison with axiomatic characterization of other solutions.

In Selten's noncooperative model the players bargain over the formation of a coalition and the division of its payoff. Players make offers, reject or accept offers in a sequential manner. In all stationary subgame-perfect equilibria of the game, if a coalition forms then it divides its worth according to a semi-stable demand vector (or with some additionally refinement concept—according to a stable demand vector).

Example 2.3. Let $N = \{1,2,3\}$ and define an NTU game (N, V) as follows:

$$\begin{aligned} V(i) &= \{0^i\},\ i \in N && (2.2)\\ V(i,j) &= \{(x^i, x^j) \in \Re_+^{(i,j)} \mid x^i + x^j \leq 2\} && (2.3)\\ V(N) &= \{(x^1, x^2, x^3) \in \Re_+^N \mid x^1 + \tfrac{1}{3}x^2 + \tfrac{1}{3}x^3 \leq 1.5\} && (2.4)\end{aligned}$$

The reader may check that there is a unique stable demand vector, namely $(x^1, x^2, x^3) = (1,1,1)$. This concept predicts the formation of a two-player coalition with equal division. Any vector of the form $(x^1, 2 - x^1, 2 - x^1)$, where $0.5 \leq x^1 \leq 1$, is semi-stable. Note that the unique stable demand vector is at one end of the semi-stable set and that the unique vector in the core of the game, namely $(0.5, 1.5, 1.5)$ is at the other end. Note also that the game is not balanced (in the usual sense of Scarf). For three person balanced games, the set of stable demand vectors coincides with the core of the game, hence these demand vectors are always feasible for the grand coalition.

The following Theorem is due to Bennett and Zame [5]:

Theorem 2.4. *Let* SSD *and* SD *be the correspondences that associate with each* NTU *game* (N, V) *the sets of semi-stable and stable demand vectors, respectively. Then, for any* NTU *game* (N, V), $\mathrm{SSD}(N, V) \neq \emptyset$ *and* $\mathrm{SD}(N, V) \neq \emptyset$.

We note that the non-levelness condition (1.1) is essential for the proofs of the statements in the previous theorem.

3. The Reduced, and Converse Reduced Game Properties

Definition 3.1. Let (N, V) be an NTU game, let $x \in \Re^N$ and let S be a coalition in N. The *reduced game* (S, V_x) is defined by:

$$V_{x,S}(T) = \bigcup_{Q \subseteq N\setminus S} \{y^T \mid (y^T, x^Q) \in V(T \cup Q)\} \qquad (3.1)$$

Interpretation: Our version of reduced games suits the purpose of analyzing demand vectors: Assume that it is agreed that members of $N \setminus S$, will get their demands $x^{N\setminus S}$. Then, members of a coalition $T \subseteq S$ may still cooperate with members of $N \setminus S$ and, provided that those players get indeed their demands, players in T can expect to obtain payoffs in $V_x(T)$.

Note that the expectations of different coalitions may not be compatible because they may require cooperation of the same subset of $N \setminus S$. The expectations expressed by the reduced game are somewhat optimistic.

Lemma 3.2. *The reduced games are games, i.e., satisfy the conditions in Definition 2.1.*

Proof: The proofs for compactness and comprehensiveness are straightforward. For non-levelness, assume that for (S, V_x) we have a coalition T in S and $z^T, y^T \in \partial V_{x,S}(T)$ with $z^T > y^T$. Then $(z^T, x^Q) \in V(T \cup Q)$ for a certain $Q \subseteq N \setminus S$. By comprehensiveness we obtain $(y^T, x^Q) \in V(T \cup Q)$. We first show that $(z^T, x^Q) \in \partial V(T \cup Q)$: If this is not true then, by non-levelness of $V(T \cup Q)$, there exists $(u^T, u^Q) \in V(T \cup Q)$ with $(u^T, u^Q) \gg (z^T, x^Q)$. By comprehensiveness $(z^T, x^Q) \in V(T \cup Q)$ and $u^T \gg z^T$. This is a contradiction to $z^T \in \partial V_{x,S}(T)$. Similarly, we obtain $(y^T, x^Q) \in \partial V(T \cup Q)$. Because $z^T > y^T$ we obtain a contradiction to the non-levelness of $V(T \cup Q)$. ∎

Definition 3.3.

(i) Let Γ be a class of NTU games. A *solution* σ on Γ is a function that assigns to each game (N, V) in Γ a subset $\sigma(N, V)$ of $\Re^N$.

(ii) A solution σ has the *reduced game property* (RGP) if for all $(N, V) \in \Gamma$, for all $z \in \sigma(N, V)$ and for all reduced games (S, V_z) it holds that $(S, V_z) \in \Gamma$ and $z^S \in \sigma(S, V_z)$.

(iii) A solution σ has the *converse reduced game property* (CRGP) if the following statement holds for all $(N, V) \in \Gamma$ and for all $x \in \Re^N$: If for all two-person reduced games (S, V_x) it is true that $(S, V_x) \in \Gamma$ and $x^S \in \sigma(S, V_x)$, then it is also true that $x \in \sigma(N, V)$.

Lemma 3.4. *The semi-stable demand correspondence has the reduced game property.*

Proof: Let (N, V) be an NTU game and let $x \in \mathrm{SSD}(N, V)$. By Lemma 3.2 we know that (S, V_x) is a well defined game for any coalition S in N. We have to show that x^S is semi-stable in (S, V_x).

(i) Maximality: Assume that $T \subseteq S$, $T \neq \emptyset$ and $y^T \in V_x(T)$ with $y^T \gg x^T$. Then $(y^T, x^Q) \in V(T \cup Q)$ for a certain $Q \subseteq N \setminus S$. By comprehensiveness $(x^T, x^Q) \in V(T \cup Q)$. By non-levelness we obtain $(x^T, x^Q) \in \operatorname{int} V(T \cup Q)$. This contradicts the maximality of x in (N, V). Hence x^S is maximal in (S, V_x).

(ii) Feasibility: Let $i \in S$. x is feasible for i in (N, V) hence there exists a coalition $R \subseteq N$, with $i \in R$ and $x^R \in V(R)$. Let $P = R \cap S$, and let $Q = R \cap S^C$. Then $x^R = (x^P, x^Q) \in V(R)$ implies that $x^P \in V_x(P)$ and because $i \in P$ we proved that x^S is feasible for i in (S, V_x). ∎

Lemma 3.5. *The stable demand correspondence has the reduced game property.*

Proof: Let (N, V) be an NTU game and let $x \in \mathrm{SD}(N, V)$. It is clear that $\mathrm{SD}(N, V) \subseteq \mathrm{SSD}(N, V)$. By Lemma 3.4 it is enough to show that x^S satisfies the independence condition in the game (S, V_x), where S is a coalition in N. Assume on the contrary that there exist $i, j \in S$ with $F_i(x^S, V_x) \subset F_j(x^S, V_x)$. Then there exists a coalition $T \subseteq S$ with $x^T \in V_x(T)$ and $j \in T$, $i \notin T$. Then there exists a coalition $Q \subseteq N \setminus S$ with $(x^T, x^Q) \in V(T \cup Q)$. Note that $j \in T \cup Q$ and that $i \notin T \cup Q$. x satisfies independence in (N, V) hence there exists a coalition $R \subseteq N$ with $i \in R$, $j \notin R$ and $x^R \in V(R)$. Let $P = R \cap S$. Note that $P \neq \emptyset$ because $i \in P$. Then $x^R = (x^P, x^{R \setminus P}) \in V(R)$ implies that $x^P \in V_x(P)$. Since $i \in P$ and $j \notin P$, we obtain a contradiction to $F_i(x, V) \subset F_j(x, V)$. ∎

Lemma 3.6. *The semi-stable demand correspondence has the converse reduced game property.*

Proof: Let (N, V) can be NTU game, let $x \in \Re^N$ and assume that $x^S \in \mathrm{SSD}(S, V_x)$ for each $S = \{i, j\} \subseteq N$. We have to show that x is semi-stable in (N, V).

Maximality: Let R be a coalition in N with $R \neq N$. We choose $i \in R$ and $j \notin R$. Assume on the contrary that $x^R \in \operatorname{int} V(R)$. Then there exists $y^R \in V(R)$ with $y^R \gg x^R$. By comprehensiveness we obtain $(y^i, x^{R \setminus i}) \in V(R)$ and this implies $y^i \in V_{x,(i,j)}(i)$. Because $y^i > x^i$ we obtain a contradiction to the maximality of (x^i, x^j) in the game $(\{i, j\}, V_x)$. Assume now that $x^N \in \operatorname{int} V(N)$, and let S be a two-person coalition in N. Then $(x^S, x^{N \setminus S}) \in \operatorname{int} V(N)$ implies $x^S \in \operatorname{int} V_{x,S}(S)$, and we obtain again a contradiction to $x^S \in \mathrm{SSD}(S, V_x)$.

Feasibility: Let S be a two person coalition in N, and let $i \in S$. By assumption x^S is feasible for i in (S, V_x), hence there exists a coalition $T \subseteq S$ with $i \in T$ and $x^T \in V_{x,S}(T)$. Then $(x^T, x^Q) \in V(T \cup Q)$ for $Q \subseteq N \setminus S$ and $i \in T \cup Q$, hence x is feasible for i in (N, V). ∎

Lemma 3.7. *The stable demand correspondence has the converse reduced game property.*

Proof: Let (N,V) be an NTU game, let $x \in \Re^N$ and assume that $x^S \in \mathrm{SD}(S,V_x)$ for all $S=\{i,j\} \subseteq N$. Because $\mathrm{SD}(S,V_x) \subseteq \mathrm{SSD}(S,V_x)$, we know by Lemma 3.6 that $x \in \mathrm{SSD}(N,V)$. It remains to show that x satisfies the independence condition in (N,V). Let $i,j \in N$, $i \neq j$ and let $S=\{i,j\}$. Assume that there exists a coalition T in N such that $i \in T$, $j \notin T$ and $x^T \in V(T)$. Then $(x^i, x^{T\setminus i}) \in V(T)$ implies $x^i \in V_{x,S}(i)$. Because x^S satisfies independence in (S,V_x) we obtain $x^j \in V_{x,S}(j)$. (Otherwise $x^S \in V_{x,S}(S)$ and j would be dependent on i, while i is not dependent on j). This implies that $(x^j, x^Q) \in V(\{j\} \cup Q)$ for $Q \subseteq N \setminus S$, hence we obtained that $R = Q \cup \{j\}$ is such that $j \in R$, $i \notin R$ and $x^R \in V(R)$. Hence it cannot be the case that $F_j(x,V) \subset F_i(x,V)$ and x satisfies independence in (N,V). ▮

Remark 3.8. We observe an interesting relation between the structure of two-person reduced games and the independence condition: Let (N,V) be an NTU game, let x be a semi-stable demand vector and let $(\{i,j\})V_x)$ be a reduced game. By Lemma 3.5 we know that x^S is semi-stable in (S,V_x) where $S=\{i,j\}$. We define $v^i_{x,S} = \sup\{y^i \mid y^i \in V_{x,S}(i)\}$ and $v^j_{x,S}$ similarly. Assume that $F_i(x,V) \subset F_j(x,V)$ — hence x is not stable. It must hold that $x^i > v^j_{x,S}$. Because x^S is feasible for i in (S,V_x) we obtain $x^S \in \partial V_{x,S}(S)$. To conclude, x^i is *the highest* demand for i in (S,V_x) that is compatible with feasibility (for j) and i needs j to obtain this. If we think that this is not reasonable because i cannot expect j to agree to such a high demand, then i should lower his demand. This is exactly the intuition behind the additional independence condition in the definition of the stable demand vectors.

4. Axiomatic Characterization of the Semi-stable and Stable Demand Correspondences

If a solution concept satisfies the RGP and CRGP then, for the characterization of this solution, it is enough to determine how this concept behaves for games with no more than two players. This behavior, together with the consistency properties, determines then uniquely the solution for general games. For the axiomatic characterization of the demand correspondences we need several definitions:

Definition 4.1. An NTU game (N,V) is called *strictly super-additive* if for all S,T coalitions in N with $S \cap T = \emptyset$, it holds that

$$V(S) \times V(T) \subset V(S \cup T) \tag{4.1}$$

Definition 4.2. Let (N, V) be an NTU game. The *core*, $C(N, V)$, of (N, V) is defined by:

$$C(N,V) = \{x \mid x \in V(N) \text{ and } x^S \notin \operatorname{int} V(S) \text{ for all } S \subseteq N\} \quad (4.2)$$

The *strict core*, $\mathrm{SC}(N, V)$, of (N, V) is defined by:

$$\mathrm{SC}(N,V) = \{x \mid x \in V(N) \text{ and } x^S \notin V(S) \text{ for all } S \subset N\} \quad (4.3)$$

Definition 4.3. A solution σ satisfies *individual rationality* (IR) if for every $x \in \sigma(N, V)$ it is true that $x^i \geq v^i$ where v^i is defined by

$$v^i = \sup \{y^i \mid y^i \in V(i)\} \quad (4.4)$$

Definition 4.4.

(i) A solution σ satisfies *unanimity* (UN) if it coincides with the core for any two-person strictly super-additive game.

(ii) A solution σ satisfies *strict unanimity* (SUN) if it coincides with the strict core for any two-person strictly super-additive game.

It is easy to check that the semi-stable demand correspondence satisfies individual rationality and unanimity, and that the stable demand correspondence satisfies individual rationality and strict unanimity. Any unanimous, individual rational solution coincides with SSD for two-person games, and any strictly unanimous, individual rational solution coincides with SD for these games. Note that $\mathrm{SD}(N, V) = \mathrm{SSD}(N, V) = \{(v^i, v^j)\}$ for a two person game (N, V) with empty core.

Theorem 4.5. *Let Γ be the class of* NTU *games. The semi-stable demand correspondence is the unique solution on Γ that satisfies* RGP, CRGP, UN, *and* IR.

Proof: It remains to prove the uniqueness part. Let $(N, V) \in \Gamma$ and let σ be a solution concept that satisfies RGP, CRGP, UN, and IR. If $|N| = 1$, we obtain $\sigma(N, V) = \mathrm{SSD}(N, V)$ by IR. Assume then that N contains at least two players. Let $x \in \sigma(N, V)$. Because σ satisfies the RGP, we obtain $x^S \in \sigma(S, V_x)$ for all $S \subseteq N$ with $|S| = 2$. By the remark above $x^S \in \mathrm{SSD}(S, V_x)$ for all such S. Because SSD satisfies the CRGP it follows that $x \in \mathrm{SSD}(N, V)$, hence $\sigma(N, V) \subseteq \mathrm{SSD}(N, V)$. Conversely, let $x \in \mathrm{SSD}(N, V)$. Because SSD satisfies the RGP we obtain $x^S \in \mathrm{SSD}(S, V_x)$ for all $S \subseteq N$ with $|S| = 2$. Hence $x^S \in \sigma(S, V_x)$ for all such S. Because σ satisfies the CRGP we conclude that $x \in \sigma(N, V)$ and $\mathrm{SSD}(N, V) \subseteq \sigma(N, V)$. ∎

Theorem 4.6. *Let Γ be the class of NTU games. The stable demand correspondence is the unique solution on Γ that satisfies* RGP, CRGP, SUN, *and* IR.

Proof: This is completely analogous to the proof of Theorem 4.5. ∎

Comment: The set of semi-stable demand vectors is usually much larger than the set of stable demand vectors (i.e., the independence condition is quite strong). It is very surprising to find out that two solutions have almost identical characterization. The only apparent difference is that the two end points of the core of a two person strictly super-additive game are semi-stable but not stable.

5. A Comparison with Other Characterizations of Solution Concepts for NTU Games

We emphasized that the demand correspondences analyzed so far are not solution concepts in the usual sense. Nevertheless the axiomatic approach allows several comparisons with consistent solutions.

The maximality condition relates the demand vectors to the core. Peleg [12] characterizes the core of NTU games by IR, RGP, and CRGP. The core satisfies unanimity for two-person games. Not surprisingly, the difference between Peleg's result and our characterization must lie in the definition of the reduced games. For the core this definition must respect the feasibility constraints for the grand coalition. Peleg's definition of a reduced game (S, V_x) agrees with ours for all coalitions $T \subset S$ but assigns to S the set

$$V_x(S) = \{y^S \mid (y^S, x^{N \setminus S}) \in V(N)\} \tag{5.1}$$

We note that the NTU games considered by Peleg must also satisfy non-levelness (at least for the grand coalition). The core is axiomatized on the class of NTU games with a non-empty core, hence Peleg's result is not a "pure" characterization.

Next we look at the egalitarian solution for NTU games (see Kalai and Samet, 1985) which is axiomatized in Hart and Mas-Colell [7]. This solution is characterized by RGP and, for two-person games, proportionality. A solution satisfies proportionality if for any two-person game $(\{1,2\}, V)$ it chooses the *Pareto optimal* vector (x^1, x^2) that satisfies $x^1 - v^1 = x^2 - v^2$ (see our Definition 4.1 for v^i). Again, the definition of the reduced games is different from ours: It is tailored for one-point solutions and, more importantly, implies that in a reduced game the players form their expectations based on cooperation with the whole complement in N. This is different from our Definition 3.1 where the players in the reduced game have the freedom to cooperate with any sub-coalition of the complement. Hart and Mas-Colell discuss the role of different formulations of reduced games in the modeling of economic problems.

Maschler and Owen [9] define a generalization of the Shapley value for NTU hyperplane games and show that this value is consistent in a weaker sense than the reduced game property. They also show that, for general NTU games, there exists no consistent value—with the definition of Hart and Mas-Colell, or the one of Peleg, for reduced games—that satisfies Pareto-optimality, Symmetry and Covariance under linear transformations of utilities. The reader should note that the egalitarian solution does not satisfy the covariance axiom.

Moldovanu [10] studies a family of solution concepts for bargaining in NTU assignment games (or two-sided markets). This family generalizes the intersection of the core and the kernel. Any member of this family is completely characterized by a choice for two-person games (say Nash, Kalai-Smorodinski, proportional, etc., and then by RGP and CRGP. The reduced games are those used by Peleg for the core. There is no consistent generalization for general NTU games. Finally we note that a version of the RGP can replace the Independence of Irrelevant Alternatives condition in the axiomatization of the Nash bargaining solution (see Thomson and Lensberg [14]).

To conclude, we observed that consistency properties play an important role in the characterization of various concepts for NTU games. The fact that extremely different outcomes can be related in such a way seems to us remarkable.

Acknowledgments: Financial support by the Deutsche Forschungsgemeinschaft through SFB 303, at the University of Bonn, is gratefully acknowledged. We thank an anonymous referee for valuable comments.

6. References

[1] W. Albers, "Core-and Kernel-variants Based on Imputations and Demand Profiles,"in: *Game Theory and Related Topics*, O. Moeschlin and D. Palaschke, eds., North-Holland, Amsterdam, 1979.

[2] R. J. Aumann and M. Maschler,"The bargaining set for cooperative games," in *Advances in Game Theory*, Princeton University Press, Princeton, New Jersey, 1964, pp. 443–476.

[3] E. Bennett E, "The Aspiration Approach to Predicting Coalition Formation and Payoff Distribution in Side-Payments Games," *International Journal of Game Theory* **12** (1983) 1–28.

[4] E. Bennett, "Endogenous vs. exogenous coalition formation," *Economie applique, tome XXXVII* (1985) 611–635.

[5] E. Bennett and W. R. Zame, "Bargaining in Cooperative Games," *International Journal of Game Theory* **17** (1988) 279–301.

[6] M. Davis and M. Maschler, "The kernel of a cooperative game," *Naval Research Logistics Quarterly* **12** (1965) 223–259.

[7] S. Hart and A. Mas-Colell A, "Potential, Value, and Consistency," *Econometrica* **57** (1989) 589–614.

[8] E. Kalai and D. Samet, "Monotonic Solutions to General Cooperative Games," *Econometrica* **53** (1985) 307–327.

[9] M. Maschler and G. Owen, "The Consistent Shapley Value for Hyperplane Games," *International Journal of Game Theory* **18** (1989) 389–409.

[10] B. Moldovanu, "Stable Bargained Equilibria for Assignment Games Without Side Payments," *International Journal of Game Theory* **19** (1990) 171–191.

[11] B. Moldovanu and E. Winter, "Core Implementation and Increasing Returns to Scale for Cooperation," Discussion Paper, Center for Rationality, Hebrew University, Jerusalem, 1992.

[12] B. Peleg, "An Axiomatization of the Core of Cooperative Games Without Side-Payments," *Journal of Mathematical Economics* **14** (1985) 203–214.

[13] R. Selten, "A Non-Cooperative Model of Characteristic Function Bargaining," in: *Essays in Game Theory and Mathematical Economics in Honor of O. Morgenstern*, V. Böhm, H. Nachtkamp, eds., Bibliographisches Institut Mannheim, Wien-Zürich, 1981.

[14] W. Thomson and T. Lensberg "Axiomatic Bargaining with a Variable Number of Agents," Cambridge University Press, Cambridge, 1989.

[15] E. Winter, "An Axiomatization of the Stable and Semi-stable Demand Vectors by the Reduced Game Property," Discussion Paper No. A-254, University of Bonn, 1989.

Essay 11

The Resale-Proof Trade of Information as a Stable Standard of Behavior – An Application of the Theory of Social Situations

Shigeo Muto[1] and Mikio Nakayama[2]

ABSTRACT. The theory of social situations due to Greenberg is applied to information trading. The solution concept, defined by Nakayama, Quintas, and Muto as a resale-proof trade, is shown to be derived as a unique stable standard of behavior for a situation which we call the information trading situation.

1. Introduction

The notion of a resale-proof trade has been introduced in [10] to formulate a trade of information under the following environments: resales of information are freely allowed and players can freely communicate and make agreements with each other; but agreements are not binding. The agreement on never reselling acquired information was called *self-enforcing* if no single player has an incentive to violate it when every other player keeps it. A resale-proof coalition is one in which the agreement can be self-enforcing. It turned out that any minimal-size resale-proof coalition attains the maximal aggregate profit over all resale-proof coalitions, and that the original seller can exploit the surplus produced by selling information to the buyers in the coalition (see [10, 9]). This is the main property of the resale-proof trade.

The same result has been obtained so far in a variety of game models. The core is adopted in the coalitional form in [10, 9]; and the bargaining

[1]Faculty of Economics, Tohoku University, Sendai 980, Japan.

[2]Faculty of Economics, Hosei University, Tama-Campus, 4342 Aihara, Machida, Tokyo 194-02, Japan.

set is considered in Muto and Nakayama [8] with appropriate modifications to that of Aumann and Maschler [1]. In a noncooperative model in Muto [7], in which trading is formulated as an extensive-form game, a modified version of his previous model [6], the coalition-proof Nash equilibrium due to Bernheim, Peleg and Whinston [2] has been applied. In the present paper, we shall characterize the resale-proof trade as a stable standard of behavior in the theory of social situations due to Greenberg [3, 4].

The self-enforcing agreement in a resale-proof trade is a condition which was exogenously imposed on the behavior of players. A behavioral basis from which it can be derived has been lacking thus far. In the extensive form in [7], this condition holds at an equilibrium; but the analysis required an additional assumption on a trading manner which was unnecessary in the coalitional-form approach. In contrast, the theory of social situations does not require any such assumptions on the behavior of players. It only requires us to describe a model in the form of *situation*. The behavior of players can then be analyzed from a single, basic standpoint, *the stable standard of behavior*. We will show that trading in a minimal-size resale-proof coalition with the self-enforcing agreement is derived as a unique stable standard of behavior for our information trading situation. In addition, one may show that it is also given as a unique vN-M stable set for an abstract system associated with our information trading situation. Thus the approach to be taken in this paper will overcome the short-coming remained in the former analyses.

There is another point in the present model of information trading. Information is assumed, as in [7, 8, 9, 10], to have the property that the profit from it is non-increasing in the number of its holders. Besides, it is assumed here that a profit to a *non-holder* is also non-increasing in the number of its holders. Since a technological innovation can be thought of as a typical example possessing this property (see, e.g., [5]), this assumption may extend the scope of our analysis. In addition, the above mentioned result in Muto [7] crucially depends on the assumption that the profit to a non-holder is invariably zero, so that his extensive-form game may not provide a basis for considering the resale-proof trade of information with the above property.

In the next section, we state assumptions and rules on the trading, define strategies and the resale-proof trade, and state a result characterizing the resale-proof trade. In section 3, the trading model is set out in the form of a social situation, and the main theorem is proved. Finally, in section 4, we conclude with several remarks including the abstract vN-M stable set.

2. The Resale-Proof Trade of Information

Let $N = \{1, 2, \ldots, n\}$ be the set of all players, where player 1 is a seller of information and players $2, \ldots, n$ are buyers. The information held by player 1 is completely replicable without costs, so that for any player it is useless to acquire more than one unit of it. Every player has a profit function depending on how many players have acquired the information. When the information is held by h players, the profit to each holder and nonholder is given by $W(h)$ and $L(h)$, respectively, in terms of money. $W(h)$ and $L(h)$ satisfy the following conditions:

Assumption 2.1.

(i) $W(h) > L(h)$ *for* $h = 1, 2, \ldots, n-1$.

(ii) $W(1) \geq W(2) \geq \cdots \geq W(n) \geq 0$ *and*
$L(1) \geq L(2) \geq \cdots \geq L(n-1) \geq 0$.

Thus the profit of each holder is uniformly greater than that of each nonholder. We assume that the values of $W(h)$ and $L(h)$ are common knowledge. We now describe trading environments. There exists no external authority that can regulate and enforce agreements on trading information. Reproduction and resales of the information are free. Players can communicate and make agreements with each other. In particular, the seller and buyers can agree, in advance, not to resell the information, but the agreement is not binding. Under such environments and Assumption 2.1, the seller may seek to make a self-enforcing agreement which maximizes his total profit. The notion of resale-proofness has been proposed to deal with this problem (see [10]). When we assume that every $i \in H$ has acquired the information, we express this as state (H). Then, at state (H), player $i \in H$ can choose buyers and a uniform price; and the designated buyers decide to buy or not at that price. Formally, local strategies at state (H) are given as follows.

Decision s_i of player $i \in H$:
$s_i \in S_i(H) := \{\emptyset\} \cup \{(T_i, p_i) | \emptyset \neq T_i \subseteq N \setminus H,\ p_i \in [0, \infty)\}$, where $s_i = \emptyset$ means that i does not sell. Put $s_H := (s_i)_{i \in H}$ and $J(s_H) := \{j \in N \setminus H | j \in T_i$ for some $s_i \neq \emptyset\}$.

Decision $b_j(s_H)$ of player $j \in J(s_H)$, given s_H with $J(s_H) \neq \emptyset$:
$b_j(s_H) \in B_j(s_H) := \{0\} \cup \{i \in H | j \in T_i\}$, where $b_j(s_H) = 0$ means that j does not buy. Put $b(s_H) := (b_j(s_H))_{j \in J(S_H)}$.

Thus, for each player $i \in N$, local strategies are defined with respect to state (H). Let $(s_H, b(s_H))$ be a local strategy combination at state (H), and define

$T(s_H, b(s_H)) := \bigcup_{i\in H^+} T_i$
where $H^+ := \{i \in H \mid s_i \neq \emptyset \text{ and } b_j(s_H) = i \text{ for all } j \in T_i\}$.

Then, H^+ is the set of those players in H whose buyers all agree to buy. When $H^+ = \emptyset$, we let $T(s_H, b(s_H)) = \emptyset$. Note that all the T_i with $i \in H^+$ are disjoint. We assume the rule that $T(s_H, b(s_H))$ is the set of new holders, that is, that any player $i \in H$, having chosen $T_i \neq \emptyset$, does not carry out the trade unless all buyers in T_i agree to buy. Buyers in T_i are assumed to buy when and only when:

$$W(|H \cup T(s_H, b(s_H))|) - p_i \geq L(|H|)$$

where $|\cdot|$ denotes the number of players in a set.

The special local strategies s_H^0 at state (H) is called the *agreement at* (H) *not to resell the information*, or the agreement, for short, if $s_i^0 = \emptyset$ for all $i \in H$. Although $b(s_H)$ is undefined at $s_H = s_H^0$, we write, for convenience, that $s_H^0 = (s_H^0, b(s_H^0))$ for all $b_j(\cdot)$ and $j \in N \setminus H$. Note that the agreement s_H^0 is tentative in nature, since it cannot be binding.

Let $u_i(s_H^0)$ denote the tentative payoff to i at state[3] H. For $i \in N \setminus H$, let $u_i(s_H^0) = L(|H|)$. We shall say the agreement s_H^0 at (H) is *self-enforcing* if no player $i \in H$ has an incentive to deviate from s_H^0 by choosing a set of buyers $T_i \neq \emptyset$ such that the agreement at state $(H \cup T_i)$ becomes self-enforcing. Formally, we define the self-enforcing agreement in a backward inductive way as follows.

Definition 2.2. Let $1 \in H \subseteq N$, and let s_H^0 be the agreement at (H). Then

(i) If $H = N$, we say s_H^0 is self-enforcing.

(ii) Suppose the definition is completed for all H with $|H| \geq h+1$. Then, for H with $|H| = h$, we say s_H^0 is self-enforcing if and only if[4]

$$u_i(s_H^0) \geq u_i(s_{H\cup T}^0) \quad \text{where } T = T(s_{H\setminus\{i\}}^0, s_i, b(s_{H\setminus\{i\}}^0, s_i)) ,$$

[3]The value of $u_i(s_H^0)$ depends upon the way how the members in H has come to share the information. For example, if $H = \{1\} \cup T(s_1, b(s_1))$ for some $s_1 = (T_1, p_1)$ then $u_i(s_H^0) = W(|H|) + |T(s_1, b(s_1))|p_1$ if $i = 1$; and $u_i(s_H^0) = W(|H|) - p_1$ if $i \in T(s_1, b(s_1))$. In general, $u_i(s_H^0) = W(|H|) + q_i$ for some $q = (q_i)_{i\in H}$ where $\sum_{i\in H} q_i = 0$. However, this insufficiency of the notation s_H^0 causes no problem in the analysis. See also footnote 4.

[4]Since $T = T(s_{H\setminus\{i\}}^0, s_i, b(s_{H\setminus\{i\}}^0, s_i))$, i is the only player to resell at state (H). Hence, $u_i(s_{H\cup T}^0) - u_i(s_H^0) = W(|H \cup T|) + |T|p_i - W(|H|)$ so that $u_i(s_H^0) \geq u_i(s_{H\cup T}^0)$ if and only if $W(|H|) \geq W(|H \cup T|) + |T|p_i$. This shows that the self-enforcingness of s_H^0 is determined independently of how payoffs have been made up to state (H).

for all $i \in H$, for all $s_i \in S_i(H)$ and for all $b(s^0_{H\setminus\{i\}}, s_i)$ such that the agreement $s^0_{H\cup T}$ at $(H \cup T)$ is self-enforcing.

This definition includes the case where s^0_1 itself is self-enforcing. This is the case where the original seller does not sell to any buyer. When s^0_1 is not self-enforcing, the definition states that there exist $s_1 = (T_1, p_1) \neq s^0_1$ and $b(s_1)$ such that for $H = \{1\} \cup T(s_1, b(s_1))$, s^0_H is self-enforcing and $u_1(s^0_1) < u_1(s^0_H)$. Note that $T(s_1, b(s_1)) = T_1$, because $T(s_1, b(s_1)) \neq \emptyset$ and our rule imply that $b_j(s_1) = 1$ for all $j \in T_1$. Then, if s^0_1 is not self-enforcing, the seller may choose such $s_1 = (T_1, p_1)$ that maximizes his payoff.

Definition 2.3. A local strategy combination $(s^*_1, b^*(s^*_1))$ at state $(\{1\})$ is a resale-proof trade if and only if for $H^* = \{1\} \cup T(s^*_1, b^*(s^*_1))$, $s^0_{H^*}$ is self-enforcing and

$$u_1(s^0_{H^*}) \geq u_1(s^0_H)$$

for all $(s_1, b(s_1))$ such that for $H = \{1\} \cup T(s_1, b(s_1))$, s^0_H is self-enforcing.

When s^0_1 is self-enforcing, the unique resale-proof trade $(s^0_1, b(s^0_1))$ is an empty trade.

We conclude this section by showing a theorem that states that in a resale- proof trade the seller chooses the minimum number of buyers necessary to form the self-enforcing agreement, and sells to them at the maximum price. Let $\mathcal{H} := \{H \subseteq N | 1 \in H \text{ and } s^0_H \text{ is self-enforcing }\}$. $\mathcal{H}$ is nonempty since $N \in \mathcal{H}$. Then:

Theorem 2.4. *Let $H^* \in \mathcal{H}$ satisfy that $|H^*| \leq |H|$ for all $H \in \mathcal{H}$. Then, a local strategy combination $(s^*_1, b^*(s^*_1))$ at state $(\{1\})$ is a resale-proof trade if and only if*

(i) $H^* = \{1\}$ *and* $s^*_1 = s^0_1$, *or*

(ii) $\{1\} \subset H^*$, $T^*_1 = H^* \setminus \{1\}$, $p^*_1 = W(|H^*|) - L(1)$, *and* $b^*_j(s^*_1) = 1$ *for all* $j \in T^*_1$.

To show this, we need the following two lemmas.

Lemma 2.5. *$H \in \mathcal{H}$ if and only if*

$$W(|H|) + |T|L(|H|) \geq (1 + |T|)W(|H \cup T|)$$

for all $T \subseteq N \setminus H$ such that $H \cup T \in \mathcal{H}$.

Lemma 2.6. *Let $H, H' \in \mathcal{H}$ satisfy that $\{1\} \subseteq H \subset H'$. Then,*

$$|H|W(|H|) - (|H| - 1)L(1) > |H'|W(|H'|) - (|H'| - 1)L(1).$$

The proofs of these lemmas will be given in the appendix. Lemma 2.5 is a re-statement of the inequality in Definition 2.2 in terms of $W(\cdot)$, $L(\cdot)$ and $p_i = W(|H \cup T|) - L(|H|)$, the maximal price that i can require at (H). Lemma 2.5 states that $H \in \mathcal{H}$ with a smaller size gives a greater maximal payoff to 1. Theorem 2.4 can be proved from these lemmas.

Proof of Theorem 2.4: Assume that $s_1^0 = s_1^*$ is self-enforcing. Then $(s_1^*, b^*(s_1^*))$ is the unique resale-proof trade. Otherwise, there exist $s_1 = (T_1, p_1) \neq s_1^0$ and $b(s_1)$ such that $H = \{1\} \cup T(s_1, b(s_1)) \in \mathcal{H}$ and that $u_1(s_1) < u_1(s_H^0)$. It suffices to show that $s_1^* = (T_1^*, p_1^*)$ gives the maximum of $u_1(s_H^0)$. Assumption 2.1 and Lemma 2.5 imply that $0 \leq W(1) - W(|\{1\} \cup T|) < |T|(W(|\{1\} \cup T|) - L(1)$ for all $\{1\} \cup T \in \mathcal{H}$. Hence, $0 \leq p_1 \leq W(|H|) - L(1)$ for all $s_1 = (T_1, p_1)$ such that $H = \{1\} \cup T(s_1, b(s_1)) \in \mathcal{H}$ because $u_1(s_H^0) - u_1(s_1) = W(|H|) - p_1 - L(1) \geq 0$ for all $i \in T(s_1, b(s_1))$. Then, Lemma 2.6 implies that for all such $s_1 = (T_1, p_1)$ we have

$$\begin{aligned}
u_1(s_H^0) &= W(|H|) + (|H| - 1)p_1 \\
&\leq W(|H|) + (|H| - 1)(W(|H|) - L(1)) \\
&= |H|W(|H|) - (|H| - 1)L(1) \\
&\leq |H^*|W(|H^*|) - (|H^*| - 1)L(1) \\
&= W(|H^*|) + (|H^*| - 1)p_1^* \\
&= u_1(s_{H^*}^0).
\end{aligned}$$

Hence, $u_1(s_H^0)$ is maximized at $s_1^* = (T_1^*, p_1^*)$, which completes the proof. ∎

3. The Stable Standard of Behavior

The resale-proof trade is defined to obtain the trade in which the payoff to the seller is maximal in all such trades that the agreement with the buyers is self-enforcing. This is an assumption we made on the behavior of the seller and buyers. In this section, we shall base our argument on the theory of social situations due to Greenberg [3] and show that the resale-proof trade can be derived from a standard of behavior. We only review below a part of the theory necessary for the later discussion. For the details of the theory, refer to Greenberg [3].

The theory of social situations first requires us to completely describe the environment as a *situation* (γ, Γ) where Γ is the set of *positions* and γ is a mapping called *the inducement correspondence.* A position $G \in \Gamma$ describes the current state of affairs in terms of a triple $(N(G), X(G), \{u_i(G)\}_{i \in N(G)})$, where $N(G)$ is the set of players, $X(G)$ is the set of all potentially feasible outcomes, and $u_i(G)$ is the utility function of player i *in position* G

over the outcomes. The inducement correspondence γ specifies, for each $G \in \Gamma$, $x \in X(\Gamma)$ and $S \subseteq N(G)$, the set $\gamma(S|G,x) \subseteq \Gamma$ of alternative positions that coalition S can induce when it rejects x in position G. Thus, a situation (γ, Γ) gives a complete specification of the social interactions.

We now describe a situation representing our information trading, called hereafter *the information trading situation*, or the IT situation for short. The IT situation is similar, at least in the spirit, to what is called in Greenberg [3] *the individual contingent threats situation.* The IT situation describes an open negotiation process, in which every single individual is free to object to a currently proposed outcome, and the outcome is adopted only when all individuals consent to accept it.

Recall that in our trading there is only one seller, player 1. Therefore, a set of all potentially feasible outcomes available to all players corresponds to the set of all agreements between player 1 and buyers. Thus, initially, we define the position:

$$\begin{aligned}
&G_N = (N(G_N), X(G_N), \{u_k(G_N)\}_{k\in N(G_N)}), \text{ where}\\
&\quad N(G_N) = N,\\
&\quad X(G_N) = \{s^0_{\{1\}\cup T} | s_1 = (T_1, p_1) \in S_1(\{1\})\}, \text{ where } T = T(s_1, b(s_1)),\\
&\quad u_k(G_N)(s^0_{\{1\}\cup T}) = u_k(s^0_{\{1\}\cup T}) \text{ for all } k \in N(G_N).
\end{aligned}$$

For each outcome $s^0_{\{1\}\cup T} \in X(G_N)$ in position G_N, every player $i \in \{1\}\cup T$ speculates if there is a resale at state $(\{1\}\cup T)$ that makes i better off. Let $(s_i|H)$ represent that at state (H) player i proposes to choose $s_i \in S_i(H)$. Then, for all $(s_i|H)$ such that

$$T(s^0_{H\setminus\{i\}}, s_i, b(s^0_{H\setminus\{i\}}, s_i)) \neq \emptyset$$

we define the position:

$$\begin{aligned}
&G_i(s_i|H) = (N(G_i(s_i|H)), X(G_i(s_i|H)), \{u_k(G_i(s_i|H))\}_{k\in N(G_i(s_i|H))}),\\
&\quad \text{where}\\
&\quad N(G_i(s_i|H)) = N,\\
&\quad X(G_i(s_i|H)) = \{s^0_{H\cup T}\}, \text{ where } T = T(s^0_{H\setminus\{i\}}, s_i, b(s^0_{H\setminus\{i\}}, s_i)),\\
&\quad u_k(G_i(s_i|H))(s^0_{H\cup T}) = u_k(s^0_{H\cup T}) \text{ for all } k \in N(G_i(s_i|H)).\ ^5
\end{aligned}$$

[5]Since $T = T(s^0_{H\setminus\{i\}}, s_i, b(s^0_{H\setminus\{i\},s_i})) \neq \emptyset$, $u_k(s^0_{H\cup T})$ is given by:

$$\begin{aligned}
&u_k(s^0_{H\cup T})\\
&\quad = u_i(s^0_H) - W(|H|) + W(|H\cup T|) + |T|p_i, \text{ for } k = i,\\
&\quad = u_k(s^0_H) - W(|H|) + W(|H\cup T|), \text{ for } k \in H\setminus\{i\},\\
&\quad = u_k(s^0_H) - L(|H|) + W(|H\cup T|) - p_i = W(|H\cup T|) - p_i, \text{ for } k \in T,\\
&\quad = u_k(s^0_H) - L(|H|) + L(|H\cup T|) = L(|H\cup T|), \text{ for } k \in N\setminus(H\cup T).
\end{aligned}$$

Thus, $G_i(s_i|H)$ is a position describing that at state (H) player $i \in H$ openly declares that he will choose $s_i \in S_i(H)$ and generate the alternative outcome $s^0_{H\cup T}$ if every other player $j \in H$ sticks to s^0_j. Note that the outcome set of every position $G_i(s_i|H)$ is a singleton, and that $G_i(s_i|H)$ is undefined for $H = N$, $s_i = s^0_i$ or $s_i = (T_i, p_i)$ with p_i large enough.

Put

$$\Gamma := \{G_N\} \cup \{G_i(s_i|H) \mid 1 \in H \subset N, i \in H, s_i \in S_i(H) \quad \text{and} \\ T(s^0_{H\setminus\{i\}}, s_i, b(s^0_{H\setminus\{i\}}, s_i)) \neq \emptyset\} .$$

Then, the inducement correspondence $\gamma \subseteq \Gamma$ is defined for all $G \in \Gamma$ as follows:

$$\begin{aligned}
&\gamma(\{j\})|G_N, s^0_{\{1\}\cup T\}}) \\
&\quad = \{G_1(s_1|\{1\})|s_1 \in S_1(\{1\}))\}\cup \\
&\qquad \{G_1(s_1|\{1\} \cup T)|s_1 \in S_1(\{1\} \cup T)\} && \text{if } j = 1, \\
&\quad = \{G_j(s_j|\{1\} \cup T)|s_j \in S_j(\{1\} \cup T)\} && \text{if } j \in T, \\
&\quad = \emptyset && \text{if } j \notin \{1\} \cup T,
\end{aligned}$$

$$\begin{aligned}
&\gamma(\{j\}|G_i(s_i|H), s^0_{H\cup T}) \\
&\qquad = \{G_j(s_j|H \cup T|)|s_j \in (H \cup T)\} && \text{if } j \in H \cup T, \\
&\qquad = \emptyset && \text{if } j \notin H \cup T,
\end{aligned}$$

and for all $G \in \Gamma$, $x \in X(G)$ and $S \subseteq N(G)$ with $|S| > 1$, we define $\gamma(S|G, x) = \emptyset$.

In the IT situation thus defined, the irreversibility inherent in information trading is embodied in the inducement correspondence γ. Namely, for any outcome $s^0_{H\cup T}$, every player j in $H \cup T$ can only induce positions $G_j(s_j|H \cup T)$ taking as given the state $(H \cup T)$. This is so because when reselling the information every player must assume that every other player in $H \cup T$ has acquired it. Only player 1, the original seller, can induce positions $G_1(s_1|\{1\})$ from G_N and revise his proposal $s^0_{\{1\}\cup T}$ made in G_N. For this reason, our IT situation is not the same to the individual contingent threats situation in Greenberg [3].

Another important distinction is that the IT situation is *hierarchical* (Greenberg [3, on p. 43, Definition 5.11]).[6] In fact, letting $\Gamma_o = \{G_N\}$ and $\Gamma_h = \{G_i(s_i|H)||H| = h\}, h = 1, ..., n-1$, one may easily show that the following two conditions characterizing the hierarchical situation are satisfied:

[6]The authors wish to acknowledge a referee for pointing out this fact. Our original proof of the main theorem (Theorem 3.1) has been considerably shortened by the use of this fact.

H.1 For each $h \in \{0, 1, ..., n-1\}$ and $G \in \Gamma_h$, $(\gamma, \{G\} \cup \Gamma_{h+1} \cup \cdots \cup \Gamma_n)$ is a situation, where $\Gamma_n := \emptyset$.

H.2 For every $G \in \Gamma$, if $G \in \gamma(S|, x)$ for some $x \in X(G)$, then S is unique in G.

We now turn to the solution theory. Let (γ, Γ) be the IT situation. For each position $G \in \Gamma$, a subset $\sigma(G)$ of $X(G)$ is called a *solution for the position* G, and a mapping σ that assigns to each $G \in \Gamma$ a solution $\sigma(G)$ is called a *standard of behavior for* Γ, or SB, for short. An SB σ for Γ is said to be *internally stable for* (γ, Γ) if for all $G \in \Gamma$, $x \in \sigma(G)$ implies that there do not exist $j \in N(G), G' \in \gamma(\{j\}|G, x)$ and $y \in \sigma(G')$ such that $u_j(G')(y) > u_j(G)(x)$; and it is said to be *externally stable for* (γ, Γ) if for all $G \in \Gamma$, $x \in X(G) \setminus \sigma(G)$ implies that there exist $j \in N(G)$, $G' \in \gamma(\{j\}|G, x)$ and $y \in \sigma(G')$ such that $u_j(G')(y) > u_j(G)(x)$. An SB σ for Γ is said to be *stable* if it is both internally and externally stable.

We are now ready to state our main result.

Theorem 3.1. *There exists a unique stable SB for the IT situation* (γ, Γ). *The stable SB* σ *is given by the following (i) and (ii).*

(i) $\sigma(G_N) = \{s^0_{H^*} \in X(G_N) | (s_1^*, b^*(s_1^*))$ *is a resale-proof trade* $\}$.

(ii) For each $G_i(s_i|H) \in \Gamma - \{G_N\}$,

$$\sigma(G_i(s_i|H)) = \begin{cases} \{s^0_{H\cup T}\} & \text{if } s^0_{H\cup T} \text{ is self-enforcing,} \\ \emptyset & \text{otherwise.} \end{cases}$$

Proof: By Theorem 5.2.1 in Greenberg [3], if there exists a stable SB for a hierarchical situation, then it is unique. Hence, it suffices to show that the SB σ given by (i) and (ii) is in fact a stable SB.

Internal stability: Let $s^0_{H^*} = s^0_{\{1\}\cup T^*} \in \sigma(G_N)$. Suppose that there exist $j \in 1 \cup T^*, G = G_j(s_j|\{1\} \cup T^*)$ and $s^0_{\{1\}\cup T^* \cup R} \in \sigma(G)$ such that $u_j(G_N)(s^0_{\{1\}\cup T^*})$, where $R = T(s^0_{(\{1\}\cup T^*)\setminus\{j\}}, s_j, b(s^0_{(\{1\}\cup T^*)\setminus\{j\}}, s_j))$. Then:

$$u_j(s^0_{\{1\}\cup T^* \cup R}) > u_j(s^0_{\{1\}\cup T^*}).$$

Since $s^0_{\{1\}\cup T^*}$ is self-enforcing by assumption, $s^0_{\{1\}\cup T^* \cup R}$ must not be self-enforcing. But this contradicts 2. If for $j = 1$ there exist $G = G_1(s_1|\{1\}) \in \gamma(\{1\}|G_N, s^0_{\{1\}\cup T^*})$ and $s^0_{\{1\}\cup Q} \in \sigma(G)$ such that

$$u_1(G)(s^0_{\{1\}\cup Q}) > u_1(G_N)(s^0_{\{1\}\cup T^*}),$$

where $Q = T(s_1, b(s_1))$, then we have

$$u_1(s^0_{\{1\}\cup Q}) > u_1(s^0_{\{1\}\cup T^*}).$$

But this contradicts the maximality of the resale-proof trade $(s_1^*, b^*(s_1^*))$.

Next, let $x \in \sigma(G_i(s_i|H))$. Then we can show in a similar way as the former part of the proof above that there exist no $j \in N(G_i(s_i|H))$, $G \in \gamma(\{j\}|G_i(s_i|H), x)$ and $y \in \sigma(G)$ such that $u_j(G)(y) > u_j(G_i(s_i|H))(x)$. This completes the proof that the SB σ is internally stable.

External stability: Let $s^0_{\{1\}\cup T} \in X(G_N)\backslash\sigma(G_N)$, where $T = T(s_1, b(s_1))$. Then $(s_1, b(s_1))$ is not a resale-proof trade. Then, either one of the following (a) and (b) must be true:

(a) $s^0_{\{1\}\cup T}$ is not self-enforcing.

(b) $s^0_{\{1\}\cup T}$ is self-enforcing, but $u_1(s^0_{\{1\}\cup T})$ is not maximal for all $s^0_{\{1\}\cup Q}$ such that $s^0_{\{1\}\cup Q}$ is self-enforcing, where $Q = T(s_1', b'(s_1'))$.

If (a) is true, then by definition there exist $j \in \{1\}\cup T$ and $s_j'' \in S_j(\{1\}\cup T)$ such that $u_j(s^0_{\{1\}\cup T\cup R}) > u_j(s^0_{\{1\}\cup T})$, where $s^0_{\{1\}\cup T\cup R}$ is self-enforcing and $\emptyset \neq R = T(s^0_{(\{1\}\cup T)\backslash\{j\}}, s_j'', b''(s^0_{(\{1\}\cup T)\backslash\{j\}}, s_j''))$. But this implies that j can induce a position $G'' = G_j(s_j''|\{1\}\cup T) \in \gamma(\{j\}|G_N, s^0_{\{1\}}\cup T)$ such that for $s^0_{\{1\}\cup T\cup R} \in \sigma(G'')$:

$$u_j(G'')(s^0_{\{1\}\cup T\cup R}) > u_j(G_N)(s^0_{\{1\}\cup T}).$$

If (b) is true, then for $j = 1$ there exists $s_1' \in S_1(\{1\})$ such that $u_1(s^0_{\{1\}\cup Q}) > u_1(s^0_{\{1\}\cup T})$ and $s^0_{\{1\}\cup Q}$ is self-enforcing, where $\emptyset \neq Q = T(s_1', b'(s_1'))$. Hence 1 can induce a position $G' = G_1(s_1'|\{1\})$ such that for $s^0_{\{1\}\cup Q} \in \sigma(G')$:

$$u_1(G')(s^0_{\{1\}\cup Q}) > u_1(G_N)(s^0_{\{1\}\cup T}).$$

Next, let $s^0_{H\cup T} \in X(G_i(s_i|H)) \setminus \sigma(G_i(s_i|H))$. Then, $s^0_{H\cup T}$ is not self-enforcing. The rest of the proof is similar to the case (a) above. Hence the SB σ is externally stable, and it is the unique stable SB. ∎

4. Concluding Remarks

As we have argued in [10], modelling information trading as a game in extensive form is not the only way to analyze it without a basis of binding agreements or contracts. Any extensive-form trading, modelled on a take-it-or-leave-it basis in particular, can easily lead to a price competition once the replicable information is sold at least to one buyer, benefitting only final buyers (see, e.g., Muto [6]). In contrast, the resale-proof trade provides a more sensible outcome at least to the seller, or the innovator under the environment where legal protection is absent or imperfect.

The theory of social situations has proved useful in locating our solution concept on a game-theoretical construct. We have shown that the resale-proof trade is derived as a unique stable standard of behavior for our information trading situation.The IT situation is only a special case of social situations; yet we believe it is a sensible application of the theory.

It should also be noted that in any social situation, its stable standard of behavior can be characterized as a vN-M abstract stable set for a system associated with the situation (Greenberg [3, on p. 38, Theorem 4.5]). The abstract system, denoted by $(D, \prec)$, can be defined for the IT situation (γ, Γ) as follows. D is the set of all pairs of possible positions and their outcomes, *i.e.*,

$$D := \{(G_N, s^0_{\{1\}\cup T_1})|s_1 \in S_1(\{1\})\} \bigcup \{(G_i(s_i|H), s^0_{H\cup T}|\{1\} \subsetneq H \subset N, i \in H, s_i \in S_i(H)\},$$

where $T_1 = T(s_1, b(s_1))$ and $T = T(s^0_{H\setminus\{i\}}, s_i, b(s^0_{H\setminus\{i\}}, s_i)) \neq \emptyset$. The domination relation $\prec$ over the set D can be given by:

$$(G, x) \prec (G', x') \text{ if and only if there exists an } i \in N \text{ such that } G' \in \gamma(\{i\}|G, x) \text{ and } u_i(G')(x') > u_i(G)(x).$$

Then, we can conclude from the above Greenberg's theorem that the resale-proof trades and self-enforcing agreements are obtained as a vN-M abstract stable set of the system $(D, \prec)$.

Finally, we conclude with a remark on Nash equilibria and the resale-proof trade. Greenberg has shown that given a game in strategic form and the ICT (individual contingent threats) situation constructed on it, if the set of Nash equilibria is nonempty, it is contained in the solution $\sigma(G^N)$ where G^N is the (grand) position in his notation [3, Theorem 7.4.1]. A similar relation holds in our IT situation. To see this, let $S^0 := \{s^0_H | u_i(s^0_H) \geq u_i(s^0_{H\cup T})$ for all $i \in H$ and $T = T(s^0_{H\setminus\{i\}}, s_i, b(s^0_{H\setminus\{i\}}, s_i))\}$. Then, we say a local strategy combination $(s_1, b(s_1))$ is a Nash equilibrium if $s^0_H \in S^0$ for $H = \{1\} \cup T(s_1, b(s_1))$; and

$$u_1(s^0_H) \geq u_1(s^0_K) \text{ for all } s^0_K \in S^0, \text{ where } K = \{1\} \cup T(s'_1, b'(s'_1)).$$

Note that $s^0_{H\cup T}$ is not restricted here to the class of self-enforcing agreements. Thus, if $(s^*_1, b(s^*_1))$ is a Nash equilibrium, it is a resale-proof trade; *i.e.*, $s^0_{H^*} \in \sigma(G_N)$ for $H^* = \{1\} \cup T(s^*_1, b(s^*_1))$. In terms of $W(\cdot)$ and $L(\cdot)$, we can show that $s^0_H \in S^0$ if and only if

$$W(|H|) + |T|L(|H|) \geq (1 + |T|)W(|H \cup T|) \text{ for all } T \subseteq N \setminus H.$$

However, this condition is rather stringent since it requires that the value $W(\cdot)$ diminish rapidly enough beyond the size $|H|$. The following numerical example illustrates the case in which the resale-proof trade is not a Nash equilibrium:

$$n = 5;\ W(1) = 17,\ W(2) = 15,\ W(3) = 13,\ W(4) = 10,\ W(5) = 8;$$
$$L(1) = 11,\ L(2) = 7,\ L(3) = 4, \text{ and } L(4) = 2.$$

It is easy to see that $W(|H|) + (5 - |H|)L(|H|) < (1 + [5 - |H|)W(5)$ for $|H| = 4$ and 3. Hence, s_H^0 is not self-enforcing because s_N^0 is self-enforcing. Then, letting $H^* = \{1, i\}$ for $i \in N \setminus \{1\}$, $s_{H^*}^0$ is self-enforcing because $W(2) + 3L(2) \geq 4W(5)$. Thus, $s_1^* = (\{i\}, W(2) - L(1))$ generates a resale-proof trade, since $W(1) + L(1) < 2W(2)$. But, this resale-proof trade is not a Nash equilibrium because $s_{H^*}^0 \notin S^0$, *i.e.*, $W(2) + tL(2) < (1+t)W(2+t)$ for $t = 1$ or 2.

Appendix

Proof of Lemma 2.5: Assume that $W(|H|)+|T|L(|H|) < (1+|T|)W(|H \cup T|)$ for some T such that $H \cup T \in \mathcal{H}$. Note that $T \neq \emptyset$. For $i \in H$, define $s_i = (T, p)$ where

$$p = [W(|H|) - L(|H|)/(1 + |T|) \geq 0.$$

Then, for each $j \in T$,

$$\begin{aligned} u_j(s_H^0 \cup T) &= W(|H \cup T|) - p \\ &= [(1 + |T|)W(|H \cup T|) - W(|H|) + L(|H|)]/(1 + |T|) \\ &> L(|H|) = u_j(H|s_H^0). \end{aligned}$$

Hence $b_j(s_{H\setminus\{i\}}^0, s_i) = i$, or $T = T(s_{H\setminus\{i\}}^0, s_i, b(s_{H\setminus\{i\}}^0, s_i))$. Then,

$$\begin{aligned} &u_i(s_H^0) - u_i(s_{H\cup T}^0) \\ &= W(|H|) - [W(|H \cup T|) + |T|p] \\ &= [(1 + |T|)(W(|H|) - W(|H \cup T|)) \\ &\qquad - |T|(W(|H|) - L(|H|))]/(1 + |T|) \\ &= [W(|H|) + |T|L(|H|) - (1 + |T|)W(|H \cup T|)]/(1 + |T|) \\ &< 0\,. \end{aligned}$$

Hence, s_H^0 is not self-enforcing, implying that $H \notin \mathcal{H}$.

Conversely, assume that $H \notin \mathcal{H}$. Then, $u_i(s_H^0) - u_i(s^0 H \cup T) < 0$ for some $i \in H$ and for some $s_i = (T, p)$ such that $H \cup T \in \mathcal{H}$. Since

$\emptyset \neq T = T(s^0_{H\setminus\{i\}}, s_i, b(s^0_{H\setminus\{i\}}, s_i))$, we have $b_j(s^0_{H\setminus\{i\}}, s_i) = i$ for all $j \in T$. But,

$$0 > u_i(s^0_H) - u_i(s^0_{H\cup T}) = W(|H|) - [W(|H \cup T|) + |T|p],$$

which implies

$$\begin{aligned} p &> [W(|H|) - W(|H \cup T|)]/|T| \\ &\geq [1 + |T|]W(|H \cup T|) - |T|L(|H|) - W(|H \cup T|)]/|T| \\ &= W(|H \cup T|) - L(|H|). \end{aligned}$$

Then, for all $j \in T$, we have

$$u_j(s^0_{H\cup T}) = W(|H \cup T|) - p < L(|H|) = u_j(s^0_H).$$

Hence $b_j(s^0_{H\setminus\{i\}}, s_i) = 0$, which is a contradiction. Therefore $H \in \mathcal{H}$. ∎

Proof of Lemma 2.6: Since $H, H' \in \mathcal{H}$ and $H \subset H'$, it follows from Lemma 2.5 that

$$W(|H|) + |H' \setminus H|L(|H|) \geq (1 + |H' \setminus H|)W(|H'|).$$

Then, by the monotonicity of $W(\cdot)$ and $L(\cdot)$, we have

$$\begin{aligned} &|H|W(|H|) - (|H| - 1)L(1) - [|H'|W(|H'|) - (|H'| - 1)L(1)] \\ &\quad = W(|H|) + (|H| - 1)W(|H|) - |H'|W(|H'|) + (|H'| - |H|)L(1) \\ &\quad \geq W(|H|) + (|H| - 1)W(|H'|) - |H'|W(|H'|) + (|H'| - |H|)L(1) \\ &\quad = W(|H|) - (1 + |H'| - |H|)W(|H'|) + (|H'| - |H|)L(1) \\ &\quad \geq -|H' \setminus H|L(|H|) + (|H'| - |H|)L(1) \geq 0 . \end{aligned}$$

All the inequalities become the equality only when $L(1) = L(|H|) = W(|H|) = W(|H'|)$. But, by Assumption 2.1, $W(|H|) > L(|H|)$ for all H such that $1 \in H \subseteq N$. Hence the inequality must be strict, and the desired inequality follows. ∎

Acknowledgments: Valuable comments from an anonymous referee are gratefully acknowledged. We also acknowledge a partial support from Tokyo Center for Economic Research.

5. References

[1] R. J. Aumann and M. Maschler, "The bargaining set for cooperative games," *Annals of Mathematics Studies* **52**, Princeton University Press, Princeton , New Jersey, 1963, pp. 443–476.

[2] B. D. Bernheim, B. Peleg, and M. D. Whinston, "Coalition-proof Nash equilibria concepts," *Journal of Economic Theory* **42** (1987) 1–12.

[3] J. Greenberg, *The theory of social situations: An alternative game-theoretic approach,* Cambridge University Press, New York, 1989.

[4] J. Greenberg, "An application of the theory of social situations to repeated games," *Journal of Economic Theory* **49** (1989) 278–293.

[5] M. L. Katz and C. Shapiro, "How to license intangible property," *QJE* **101** (1986) 567–589.

[6] S. Muto, "An information good market with symmetric externalities," *Econometrica* **54** (1986) 295–312.

[7] S. Muto, "Resale-proofness and coalition-proof Nash equilibria," *Games and Economic Behavior* **2** (1990) 337–361.

[8] S. Muto and M. Nakayama, "Sequential bargaining in information trading," Working paper No. 95, New York University, 1990.

[9] M. Nakayama and L. Quintas, "Stable payoffs in resale-proof trades of information," *Games and Economic Behavior* **3** (1991) 339–349.

[10] M. Nakayama, L. Quintas, and S. Muto, "Resale-proof trades of information," *The Economic Studies Quarterly* **42** (1991) 292–302.

Essay 12

The Non-Consistency and Non-Uniqueness of the Consistent Value

Guillermo Owen[1]

1. Introduction

As is well known, Shapley's value [7] can be obtained in terms of marginal payoffs to players in randomly formed coalitions. Moreover, Hart and Mas-Colell [2] have shown that the value satisfies a consistency property in terms of reduced games (see also Peleg [6]).

Maschler and Owen [4] developed a *consistent value* for hyperplane games. This consistent value is a direct generalization of the Shapley value in terms of marginal payoffs; moreover, it possesses a (slightly weaker) version of the consistency property.

Maschler and Owen [5] give a generalization of this last value for general NTU games. Under certain regularity assumptions for the sets $V(S)$, existence (though not uniqueness) of this value is proved there. Since then, Hart and Mas-Colell [3] have proved that this value is also the solution of an n-person bargaining problem. Thus, this (Maschler/Owen/Hart/Mas-Colell) value has many interesting properties. Consistency and uniqueness of this value remained open questions. In this brief note, we answer both these question in the negative.

2. The hyperplane and consistent values

Let V be an n-person NTU game (so that $V(S)$ is a subset of R^S). We assume that these sets $V(S)$ have smooth Pareto-optimal boundaries; moreover, if $\boldsymbol{x}$ is a Pareto-optimal point of $V(S)$, we let $H(\boldsymbol{x}; S)$ be the hyperplane which is tangent to $V(S)$ at point $\boldsymbol{x}$. Moreover, for any $K \subset N$, we

[1]Naval Postgraduate School, Monterey, CA 93953

define the restricted game $V|_K$ as a game with player set K, such that $V|_K(T) = V(T)$ if $T \subseteq K$, but is undefined otherwise.

Let Φ be a solution concept for games. For an n-person game V, with player set N, and some coalition $S \subseteq N$, the *reduced game* V_S^* is a game with player set S, defined by

$$V_S^*(T) = \{(x_j)_{j\in T} | ((x_j)_{j\in T}, (y_k)_{k\in N\setminus S}) \in V(N \setminus S \cup T)\} \tag{2.1}$$

(where $y_k = \Phi_k[V|_{N\setminus S\cup T}]$ for each $T \subseteq S$).

The reduced game is treated by Davis and Maschler [1] (see also Peleg [6]) although the definition there is not quite the same as ours. Heuristically, the reduced game tries to represent the bargaining that goes on within coalition S to determine how its share of the winnings from the game should be distributed among its members. Each $T \subseteq S$ then considers what it could hope to gain in a game with the members of $N \setminus S$, assuming reasonable behavior by $N \setminus S$ (reasonable being defined in terms of the solution concept Φ, applied to the restricted game $V|_{N\setminus S\cup T}$).

Now, the solution concept Φ is said to be *strongly consistent* if for every game (V, N), any player $i \in N$, and any coalition K with $i \in K \subset N$,

$$\Phi_i[V|_K] = \Phi_i[V] \ . \tag{2.2}$$

Hart and Mas-Colell [2] prove consistency of the Shapley value (for TU games). Maschler and Owen [4], however, prove that, for NTU games, consistency is incompatible with the usual axioms (symmetry, Pareto-optimality, and invariance under linear transformations of utility). They therefore define the concept of k-consistency: a solution concept Φ is *k-consistent* if for every game (V, N), where $n \geq k$, and any $i \in N$,

$$\binom{n-1}{k-1}\Phi_i[V] = \sum_{\substack{K \subset N \\ |K| = k \\ i \in K}} \Phi_i[V|_K] \ . \tag{2.3}$$

For hyperplane games (*i.e.*, those in which each set $V(S)$ is defined by a single linear inequality, so that the Pareto-optimal boundary is a hyperplane), Maschler and Owen [4] define a *random order value* as follows. Let π be a permutation (ordering) of the players. Let $P(i, \pi)$ be the set of all j such that $\pi(j) < \pi(i)$. Then, define the vector $\boldsymbol{y}^\pi$ inductively by

$$y_i^\pi = \max\left\{x_i | (x_i, (y_j^\pi)_{j\in P(i,\pi)}) \in V(P(i, \pi) \cup \{i\})\right\} . \tag{2.4}$$

The random order value is then the arithmetic mean of the vectors $\boldsymbol{y}^\pi$, *i.e.*,

$$\Phi_i = \frac{1}{n!}\sum_\pi y_i^\pi \ . \tag{2.5}$$

It is easy to see, of course, that for TU games, this random order value coincides with the Shapley value. Moreover, it can be shown that it is k-consistent for every k, over the space of hyperplane games. In fact, it is the only solution concept which is symmetric, efficient, k-consistent, and invariant under linear transformations of utility, over the space of hyperplane games. Over this space, we call it the *hyperplane value.*

Unfortunately, the random order value is not satisfactory for general NTU games. One reason for this is that, for hyperplane games, the arithmetic mean of Pareto-optimal points is also Pareto-optimal. For general NTU games, this is no longer true (it will generally be sub-optimal if the sets $V(S)$ are convex, and need not even be feasible if the $V(S)$ are not convex). Thus Maschler and Owen [5] define yet another value, the *consistent value* for TU games which, for hyperplane games, coincides with the random order value.

Let (V, N) be an NTU game. A *consistent solution* for V is a $(2^n - 1)$-tuple of points $\boldsymbol{z}^S$ (one for each coalition) such that for each S, $\boldsymbol{z}^S$ is the hyperplane value of the s-person hyperplane game determined by the $2^s - 1$ hyperplanes $H(\boldsymbol{z}^T; T)$, $T \subset S$. Existence of this solution is proved by a fixed-point argument. This solution has some interesting properties: Hart and Mas-Colell [3] show, "inter alia", that it is a subgame-perfect equilibrium solution to an n-person bargaining problem. (The reader is referred to the literature for these properties.)

Unfortunately, both [5] and [3] leave open the question of consistency of the consistent solution. Moreover, both articles use fixed-point arguments to prove the existence of the solution, and hence the question of uniqueness is also left open. We address here the two questions:

(a) is the consistent solution k-consistent?

(b) is the consistent solution unique (perhaps restricting ourselves to the case of superadditive games and convex sets $V(S)$)?

The answer, unfortunately (but not too surprisingly), is in the negative in both cases.

Example 2.1. Consider the 3-person game defined by:

$$\begin{aligned}
V(\{i\}) &= \{x_i \le 0\} \text{ for } i = 1, 2, 3. \\
V(\{1,2\}) &= \{(x_1, x_2) \mid x_1 + 2x_2 \le 2\} \\
V(\{1,3\}) &= \{(x_1, x_3) \mid 2x_1 + x_3 \le 2\} \\
V(\{2,3\}) &= \{(x_2, x_3) \mid x_2 + 2x_3 \le 2\} \\
V(\{1,2,3\}) &= \{(x_1, x_2, x_3) \mid x_1 + x_2 + x_3 \le 3\ ,\ 5x_1 + x_2 + x_3 \le 7.2\}\ .
\end{aligned}$$

Thus, all the Pareto-optimal surfaces for $V(S)$, except for $V(N)$ itself, are hyperplanes. It is not too difficult to see that the consistent solution is made up of the points

$$\begin{aligned} z^{\{i\}} &= 0 \quad \text{for each} \quad i \in N \\ z^{\{1,2\}} &= (1, \tfrac{1}{2}, -) \\ z^{\{1,3\}} &= (\tfrac{1}{2}, -, 1) \\ z^{\{2,3\}} &= (-, 1, \tfrac{1}{2}) \\ z^{N} &= (1, 1, 1). \end{aligned}$$

In fact, for $|S| = 1$, this is merely the single efficient point of $V(S)$. For $|S| = 2$, the restricted games are hyperplane games, and so the z^S are simply the random order values. For set N, we note that $(1, 1, 1)$ is Pareto-optimal in $V(N)$, and that the tangent hyperplane there is simply $x_1 + x_2 + x_3 = 3$. With this hyperplane replacing the true set $V(N)$, the game becomes symmetric (it has cyclic symmetry under the permutation $1 \to 2 \to 3 \to 1$), and so the random order value for N is $(1, 1, 1)$.

To check our solution for 2-consistency, we now compute the reduced 2-person games. For $V^*_{\{1,2\}}$, we note that, since $z_3^N = 1$,

$$V^*_{\{1,2\}}(\{1,2\}) = \{(x_1, x_2) \mid x_1 + x_2 \le 2 \ , \ 5x_1 + x_2 \le 6.2\} \ .$$

Also, since $z_3^{\{1,3\}} = 1$,

$$V^*_{\{1,2\}}(\{1\}) = \{2x_1 \le 1\}$$

and, since $z_3^{\{2,3\}} = \frac{1}{2}$,

$$V^*_{\{1,2\}}(\{2\}) = \{x_2 \le 1\} \ .$$

It is not difficult to see from this that the solution to the game $V^*_{\{1,2\}}$ is

$$\begin{aligned} z^{\{1\}} &= 0.5 \\ z^{\{2\}} &= 1 \\ z^{\{1,2\}} &= (0.75, 1.25) \ . \end{aligned}$$

The first two of these are easy; the third, $z^{\{1,2\}}$ is verified by noticing that $(0.75, 1.25)$ is Pareto-optimal for $V^*_{\{1,2\}}(\{1,2\})$, and that, at that point, the tangent line is simply $x_1 + x_2 = 2$. Thus this reduces to a TU game.

In a similar fashion, we can compute $V^*_{\{2,3\}}$, and find that it is given by

$$\begin{aligned} V^*(\{2\}) &= \{2x_2 \le 1\} \\ V^*(\{3\}) &= \{x_3 \le 1\} \\ V^*(\{2,3\}) &= \{(x_2, x_3) \mid x_2 + x_3 \le 2 \ , \ x_2 + x_3 \le 2.2\} \end{aligned}$$

(Note that this last inequality is redundant.) This is actually a TU game, with consistent value

$$\begin{aligned} z^{\{2\}} &= 0.5 \\ z^{\{3\}} &= 1 \\ z^{\{2,3\}} &= (0.75, 1.25)\ . \end{aligned}$$

Finally, $V^*_{\{1,3\}}$ is given by

$$\begin{aligned} V^*(\{1\}) &= \{x_1 \leq 1\} \\ V^*(\{3\}) &= \{2x_3 \leq 1\} \\ V^*(\{1,3\}) &= \{(x_1, x_3) \mid x_1 + x_3 \leq 2\ ,\ 5x_1 + x_3 \leq 6.2\}\ . \end{aligned}$$

For this game, the consistent value turns out to be

$$\begin{aligned} z^{\{1\}} &= 1 \\ z^{\{3\}} &= 0.5 \\ z^{\{1,3\}} &= (1.05, 0.95)\ . \end{aligned}$$

Thus, in the two 2-person reduced games to which 1 belongs, player 1 receives 0.75 and 1.05 respectively, for an arithmetic mean of 0.90. In the global game, however, 1's value is 1. This contradicts the 2-consistency relation, and we see that this value is not 2-consistent.

Next there is the problem of uniqueness. Of course it is easy to give examples with non-unique solutions by simply choosing irregular enough games. Trivially, we can let V be a 2-person game which assigns 0 to each player, and let $V(\{1,2\})$ be the set satisfying $x_1x_2 \leq 1$ in the first quadrant. This is of course not a convex set. Alternatively, we can let $V(\{1,2\})$ be the set satisfying $x_1x_2 \geq 1$ in the third quadrant. This is a convex set, but the game is not superadditive. We will therefore give an example which is both convex and superadditive.

Example 2.2. Let V be a 3-person game. We let $V(\{i\})$ here be the set $\{x_i \leq 0\}$, as in the previous example. For each pair $\{i, j\}$ of distinct players, we let

$$W(\{i,j\}) = \{(x_i, x_j) \mid x_i + x_j \leq 2\ ,\ x_i \geq 0\ ,\ x_j \geq 0\}$$

and let $V(\{i,j\})$ be the comprehensive closure of W, *i.e.*, the set of all points which are component-wise smaller than or equal to some point in W.

To define $V(\{1,2,3\})$, now, let K be the set of points satisfying

$$\begin{aligned} (1-x_1)(1-x_2)(1-x_3) &\geq .001 \\ 0 \leq x_i &\leq 1 \quad \text{for}\ \ i = 1,2,3 \end{aligned}$$

and

$$J = \{(x_1, x_2, x_3) \mid x_1 + x_2 + x_3 \leq 2\ ,\ (\forall i)(x_i \geq 0)\}\ .$$

Now, let E be the convex hull of $J \cup K$, and let $V(N)$ be the comprehensive closure of E.

It is easy to see that $V(N)$, defined this way, is closed and convex. Moreover, V is superadditive as J specifically includes each set $V(\{i,j\})+V(\{k\})$ (where i, j, and k are the three players in some order).

Now, the Pareto-optimal surfaces of the sets $V(\{i,j\})$ are all hyperplanes, and so there is no difficulty in seeing that the consistent solution will include the points

$$\begin{aligned} \boldsymbol{z}^{\{i\}} &= 0 \quad \text{for each } i \\ \boldsymbol{z}^{\{1,2\}} &= (1, 1, -) \\ \boldsymbol{z}^{\{1,3\}} &= (1, -, 1) \\ \boldsymbol{z}^{\{2,3\}} &= (-, 1, 1). \end{aligned}$$

It remains to determine the point $\boldsymbol{z}^N$. Let $\boldsymbol{y}$ be any point of K which is Pareto-optimal in $V(N)$. We claim $\boldsymbol{z}^N = \boldsymbol{y}$ satisfies the conditions of the consistent solution.

In fact, since K is smooth (the partial derivatives of the defining function are all continuous and non-zero), the hyperplane $H(\boldsymbol{y}, N)$ must be the tangent plane to K at $\boldsymbol{y}$. This has the equation

$$(1-y_2)(1-y_3)x_1 + (1-y_1)(1-y_3)x_2 + (1-y_1)(1-y_2)x_3 = c \qquad (2.6)$$

where c is equal to the left side of 2.6, with x_j replaced by y_j. This plane passes, in particular, through the points

$$\boldsymbol{p}^1 = \left(0, 2, y_3 + \frac{(y_2-2)(1-y_3)}{1-y_2} + \frac{y_1(1-y_3)}{1-y_1}\right)$$

and

$$\boldsymbol{p}^2 = \left(2, 0, y_3 + \frac{(y_1-2)(1-y_3)}{1-y_1} + \frac{y_2(1-y_3)}{1-y_2}\right)\ .$$

Some calculation shows us that the sum of the third components of these two points is

$$p_3^1 + p_3^2 = 6y_3 - 4\ . \qquad (2.7)$$

Let us see, now, what player 3 will receive, according to the hyperplane value, assuming the hyperplane $H(\boldsymbol{y}, N)$, and the hyperplanes already fixed for the smaller coalitions. We consider each permutation, as well as the amount 3 will receive in each case:

Permutation	Payoff to 3
1, 2, 3	p_3^2
2, 1, 3	p_3^1
1, 3, 2	2
2, 3, 1	2
3, 1, 2	0
3, 2, 1	0

By (8), the sum of these six payoffs is $6y_3$. Thus the mean is y_3, and we conclude that, in this hyperplane game, 3 would receive payoff y_3. Given the symmetry of the game, players 1 and 2 would receive the amounts y_1 and y_2, respectively. Hence $\boldsymbol{y}$ (together with the other $\boldsymbol{z}^S$) is a consistent value for V.

We show, finally, that there is more than one such point $\boldsymbol{y}$. In fact, J and K are both compact; so then is their union. E, as the convex hull of a compact set, is compact and convex, and is therefore the convex hull of its extreme points. But every extreme point of E must be in either J or K. Now, the only extreme points which come from J are the extreme points of J itself, namely $(0,0,0)$, $(2,0,0)$, $(0,2,0)$, and $(0,0,2)$.

Suppose now that $\boldsymbol{y}$ is an extreme point of E which comes from the Pareto-optimal surface of K. By the symmetry of the game, any permutation of the components of $\boldsymbol{y}$ is also such a point. Hence, if there is only one such point, it must be the one Pareto-optimal point of K which has three equal components, namely $(0.9, 0.9, 0.9)$.

Suppose, then, that this is the only extreme point that comes from K. Then E must be the convex hull of the five points $(0,0,0)$, $(2,0,0)$, $(0,2,0)$, $(0,0,2)$, and $(0.9, 0.9, 0.9)$. This convex hull is an irregular hexahedron, determined by the inequalities

$$9x_1 + 9x_2 + 2x_3 \leq 18 \tag{2.8}$$

$$9x_1 + 2x_2 + 9x_3 \leq 18 \tag{2.9}$$

$$2x_1 + 9x_2 + 9x_3 \leq 18 \tag{2.10}$$

$$x_1, x_2, x_3 \geq 0\ . \tag{2.11}$$

However, we note that the point $(0.8, 0.9, 0.95)$, which belongs to K, fails to satisfy inequality 2.11. Hence E must contain at least three extreme points from K. As seen above, each one of these will give a consistent solution to the game. We conclude that the consistent value is not unique.

Acknowledgments: The author gratefully acknowledges financial support for this research from the University of Bielefeld, Germany.

3. References

[1] M. Davis and M. Maschler, "The kernel of a cooperative game, *Naval Research Logistics Quarterly* **12** (1965) 223–259.

[2] S. Hart and A. Mas-Colell, "The potential of the Shapley value, in: *The Shapley value* A. E. Roth, Ed., Cambridge University Press, 1988, pp. 127–138.

[3] S. Hart and A. Mas-Colell, "A model of n-person non-cooperative bargaining, unpublished manuscript, July 1992.

[4] M. Maschler and G. Owen, "The consistent Shapley value for hyperplane games, *International Journal of Game Theory* **18** (1989) 389–407.

[5] M. Maschler and G. Owen, "A consistent Shapley value for games without side payments, in: *Rational Interaction: Essays in Honor of J. C. Harsanyi*, R. Selten, Ed., Springer-Verlag 1992, pp. 35–43.

[6] B. Peleg, "On the reduced game property and its converse, *International Journal of Game Theory* **15** (1986) 187–200.

[7] L. S. Shapley, "A value for n-person games, in: *Contributions to the Theory of Games* **II**, A. W. Tucker and H. W. Kuhn, Eds., Princeton University Press, 1953, pp. 307–317.

Essay 13

The Kernel of Homogeneous Games with Steps

B. Peleg[1], J. Rosenmüller[2], and P. Sudhölter[3]

ABSTRACT. The class of homogeneous n-person constant sum games was introduced by von Neumann-Morgenstern [20]; as a solution concept they treated the v.N.-M.-solution (or main simple solution). Peleg [9] discussed the kernel and nucleolus within this framework. The general theory of non-constant sum homogeneous games was developed by Ostmann [8], Rosenmüller [14], [15], [16] and Sudhölter [19]. Based on these results, Peleg-Rosenmüller [12] discussed the kernel (and other solution concepts) for homogeneous games without steps while Rosenmüller-Sudhölter [18] proved a reduction theorem for the nucleolus. The present paper provides the reduction theorem for the kernel: it is shown that the kernel of a homogeneous game with steps equals the one of the game without steps obtained by reducing in a suitable way, that is, cutting off all players following the largest step.

1. Introduction, notations; homogeneous games

This section is of an introductory nature. We will present our notation as well as a short survey providing the necessary foundations from the theory of homogeneous games.

Sections 2 to 5 are organized as follows. Section 2 introduces the notations for treating the kernel and its relatives. Essentially we are dealing with the pre-kernel of a directed game and, by showing that its elements are nonnegative, we eventually convince ourselves that dealing with the kernel is sufficient in our present context.

---

[1]Department of Mathematics and Center for Rationality and Interactive Decision Theory, The Hebrew University of Jerusalem, Givat-Ram, 91904 Jerusalem, Israel

[2]Institute of Mathematical Economics, University of Bielefeld, Postfach 8640, 4800 Bielefeld 1, Germany

[3]Same address as the second author

In Section 3 we concentrate on a small problem that in most papers concerning the kernel (or the nucleolus for that matter) is not explicitly mentioned: how to treat the case that winning players are present. As one can expect, this case can be eliminated in a straightforward way; however, it takes some meticulous arguing. For the corresponding treatment with respect to the nucleolus the reader is referred to Rosenmüller-Sudhölter [18]. Now, as the existence of inevitable or veto-players can be excluded in a rather easy manner, the following sections deal only with the case that neither winning players nor veto-players, nor dummies are present.

Section 4 actually presents the main burden of proofs: the technique of reducing the kernel. More precisely, it is shown that in homogeneous games (with the above mentioned restrictions) every element of the kernel has zero coordinates for all players following the largest (first) step.

Finally, Section 5 collects the results and presents the main theorem. First we prove the converse of the main theorem of Section 4, that is, the kernel of the game obtained by properly reducing or truncating the game after the first step contains elements which, augmented by zero coordinates, yield kernel elements of the original game. Hence we come up with the final result that the kernel of a game with steps equals the kernel of the reduced game.

To come back to the purpose of the present section, let us first introduce some notation and then focus on the theory of homogeneous games.

The term homogeneous game goes back to von Neumann and Morgenstern [20]; they were dealing with constant sum games only and the main purpose to introduce the topic seems to be the fact that homogeneous simple games have a nice von Neumann-Morgenstern-solution (the so called "main simple solution").

In a series of papers Isbell [2] ,[3], [4] treated several aspects of homogeneous games and later, Peleg [9], [10] studied the kernel and the nucleolus for this particular class of games. However, all these authors were concerned with constant sum games only. In 1975 one of the present authors noticed that homogeneity is a useful concept to describe extreme (convex or superadditive) games (cf. Rosenmüller [13]) and subsequently he tried to obtain an overview over the structure of a set of weights admissible for constructing homogeneous games (cf. Rosenmüller [14], [15]). However, it was Ostmann [8] who first found out that a homogeneous game has a unique minimal representation. An alternative definition was given by Rosenmüller [16] (this works also in the case of countably many players cf. [17]). In the next step Sudhölter [19] introduced the theory of the "incidence vector" showing that a homogeneous game is equivalent to an integer vector describing the minimal lengths of a certain class of minimal winning coalitions.

This series of papers was dealing with a general and nonconstant sum case which, as it turned out, was much richer than the constant sum structure. In particular, it contained the unique representation theorem of Von Neumann-Morgenstern as a simple case of Ostmann's general existence theorem for the minimal representation. So far, however, these papers were only concerned with describing the structure of homogeneous games.

The first attempt to capitalize on the new representation theory with respect to elaborating solution concepts was made by Peleg-Rosenmüller [12]. For the general homogeneous weighted majority game "without steps" they discuss the least core, the nucleolus, and the kernel and show their close relationship with the unique minimal representation. At this stage it was clear that all attempts in this direction would necessarily call for some version of a reduction theorem: as Sudhölter [19] has shown, homogeneous games in general and homogeneous games without steps can be brought into a one to one correspondence by adding just one player. On the other hand, the fact that steps play a very dominant role in the homogeneous game leads one to believe that concepts like the nucleolus and the kernel, which react very sensitively to the fact that "steps rule their followers", should somehow give a considerable advantage to players preceding the steps of a game.

Indeed, it turned out that the nucleolus vanishes after the first step and hence a computation of the nucleolus of an arbitrary homogeneous game can be reduced to the computation of the nucleolus of the reduced game which is obtained by cutting off the players after the first step [18]).

The present paper is designed to show the same property for the kernel. Eventually we will come up with the result that the kernel of a homogeneous game with steps is essentially computed by the kernel of the homogeneous game without steps which is obtained by truncating the players following the first step. Note that the techniques of the present paper and those presented in [18] are quite different. In [18] it is shown that the system of minimal winning coalitions of a homogeneous game with steps cannot be weakly balanced and then provide an induction argument for the vanishing of nucleolus coordinates after the first step. This procedure rests heavily on Sudhölter's theory of the incidence vector ([19]).

The present paper also rests on arguments implicitly producing the "incidence vector". But we do not make use of the balancedness of the system of minimal winning coalitions (see the results of [12] for the case of games without steps). Instead it is seen that sufficiently many minimal winning coalitions with maximal excess can be produced in order to determine the shape of the kernel after the largest step. Thus, the proofs provided in [18] and in the present version are completely independent; nevertheless it is of course clear that our present results induce that the nucleolus has coordinates vanishing after the first step of a homogeneous game.

We now start introducing our notation. Let $\mathcal{N} = \{1, 2, 3, ...\}$ denote the universe of *players*. Subsets of $\mathcal{N}$ are called *coalitions*; *intervals* are special types of coalitions of the form

$$[a, b] = \{i \in \mathcal{N} | a \leq i \leq b\}, \tag{1.1}$$

where $a, b \in \mathcal{N}$. The *grand coalition* is some distinguished interval Ω, predominantly we use $\Omega = [1, n]$. Next,

$$P = P(\Omega) = \{S | S \subseteq \Omega\} \tag{1.2}$$

and, if

$$v : P \to \Re, v(\emptyset) = 0 \tag{1.3}$$

is a mapping (the *coalition* (or chracteristic) *function*) then (Ω, P, v) is a *game*.

In most cases the nature of Ω and P is clear, so we call v a *game* as well. v is *simple* if $v : P \to \{0, 1\}$ holds true.

In order to simplify the notation, we write $S \cup i$ instead of $S \cup \{i\}$ etc. for $S \in P$ and $i \in \Omega$. $S + T$ and $S + i$ denotes *disjoint* unions (i.e. $S + T = S \cup T$ iff $S \cap T = \emptyset$ etc.). Given a game, the *desirability relation* is a binary relation on players (only, it can be considered on coalitions as well). Player $i \in \Omega$ is more *desirable* than player $j \in \Omega$ (written $i \succsim j$) iff

$$v(S + i) \geq v(S + j) \tag{1.4}$$

for all $S \in P$, $S \cap \{i, j\} = \emptyset$. If $i \sim j$ (i.e. $i \succsim j$ and $j \succsim i$), then i and j are *equivalent* or *of the same type*.

A game is *ordered* if $\succsim$ is complete and *directed* if the ordering coincides with enumeration, i.e., if

$$i < j \text{ implies } i \succsim j. \tag{1.5}$$

Thus, in a directed game the "strong", "more desirable", or "large" players are first in enumeration (or "index"). If a game is obviously ordered, we will always tacitly assume that it is directed since this can be enforced by just renaming the players. We write

$$\ell(S) = \max\{i | i \in S\} \tag{1.6}$$

for $S \in P$, sometimes calling this the *length* of coalition S (counting starts at $1 \in \mathcal{N}$!)—but note that in a directed game, $\ell(S)$ is also the "weakest", "smallest", or "last", player of coalition S.

Let us now turn to simple games.

Given a simple game v, $W = W(v) = \{S \in P | v(S) = 1\}$ is the system of *winning* coalitions while

$$W^m = W^m(v) = \{S \in W | v(T) = 0 \text{ for } T \subsetneqq S\} \tag{1.7}$$

is the system of *minimal* winning coalitions (*"min-win coalitions"*).

A vector $m = (m_i)_{i \in \Omega} \in \Re_+^\Omega$ is tantamount to a function on P via $m(S) = \sum_{i \in S} m_i (S \in P)$ (thus, it is a non-simple "game") and hence called a *"measure"* (m is additive). Games and in particular measures, may be restricted on subsets $T \subseteq \Omega$, the notation is $v_{|T}$ or $m_{|T}$; e.g.

$$\begin{array}{ll} v_{|T}(S) = v(T \cap S) & (S \in P(\Omega)) \\ v_{|T}(S) = v(S) & (S \in P(T)); \end{array} \tag{1.8}$$

the version living on $P(\Omega)$ and the one living on $P(T)$ are not distinguished. We tolerate $v_{|\emptyset}$.

If m is a measure and $\lambda > 0$, then $(m; \lambda)$ is a *representation* of v if

$$v(S) = \begin{cases} 1 & m(S) \geq \lambda \\ 0 & m(S) < \lambda \end{cases} \tag{1.9}$$

holds true, in this case we write $v = v_\lambda^m$. Of course, integer representations are of particular interest.

A measure m is said to be *homogeneous* w.r.t. $\lambda \in \Re_{++}$ (written "m hom λ"), if, for any $T \in P$ with $m(T) > \lambda$, there is $S \subseteq T$ with $m(S) = \lambda$.

A game v is *homogeneous* if there exists a representation $(m; \lambda)$ with m hom λ and $v(\Omega) = 1$. (The definition is due to von Neumann–Morgenstern [20].)

Clearly, a representable game is always ordered, thus in accordance with our remark above we tacitly assume that representable games are directed. That is, there exists a representation $(m; \lambda)$ such that $i < j$ implies $m_i \geq m_j (i, j \in \Omega)$.

While players are ordered according to "size", coalitions are ordered lexicographically. In particular, the *lex-max min-win coalition* is the lexicographically first minimal winning coalition; in a homogeneous game with homogeneous representation $(m; \lambda)$ this coalition is denoted by $S^{(\lambda)}$ (an interval with measure $m(S^{(\lambda)}) = \lambda$). We write $\ell^{(\lambda)} = \ell(S^{(\lambda)})$.

Player $i \in \Omega$ is a *dummy* if $v(S \cup i) = v(S)$ for all $S \in P$; this notion is not restricted to simple games. All dummies are equivalent or of the same type.

Returning to simple games, player $i \in \Omega$ is *winning* if $v(\{i\}) = 1$ (i.e. $\{i\} \in W^m$); again all winning players are equivalent.

Thus dummies and winning players provide two particular types; of course types establish a decomposition of Ω.

There is a second decomposition of Ω which is fundamental in the case of a homogeneous game, this is the decomposition into sets of players of equal *character*. To explain the nature of this notion, let us focus on a homogeneous game.

There are three characters to be attached to players w.r.t. a homogeneous game called "dummy", "sum", and "step". The dummy character is the dummy type as explained above, so let us attempt to define the other two. To this end, fix a non-dummy player $i \in \Omega = [1, n]$ and consider the minimal length of all min-win coalitions containing i, say

$$\ell^{(i)} := \min\{\ell(S) | S \ni i, S \in W^m\}. \tag{1.10}$$

The *domain* of i is

$$C^{(i)} := [\ell^{(i)} + 1, n]. \tag{1.11}$$

Now, player i is a sum ("his character is sum") if

$$m_i \leq m(C^{(i)}), \tag{1.12}$$

otherwise i is a *step*.

A sum may be replaced in at least one min-win coalition by a coalition of smaller players, his weight being exactly the sum of the weights of the smaller players (homogeneity is essential here).

On the other hand, "steps rule their followers", i.e., whenever a smaller player is a member of a min-win coalition, any preceding step is also a member.

Thus Ω decomposes into three subsets of characters: dummy, sum, and step. (Note that a winning player may be sum or step). A game may have no dummies or sums (e.g. the unanimous game of the grand coalition) but steps are always present.

The following remark collects some facts from the theory of homogeneous games. The details can be found e.g. in Ostmann [8], Rosenmüller [16] and Sudhölter [19].

Remark 1.1.

1. The smallest non-dummy player is always a step. If v is constant sum, then this is the only step. To simplify matters, we say v is a (homogeneous) *game without steps* if the smallest non-dummy is the only step.

2. A homogeneous game v has a unique minimal representation. (i.e., an integer valued $(\bar{m}; \bar{\lambda})$, representing v such that $\bar{m}(\Omega)$ is minimal among all integer representation—this $(\bar{m}; \bar{\lambda})$ is homogeneous.) For games without steps and dummies, every representation is a multiple of the minimal one.

3. Again let $S^{(\lambda)}$ be the lex-max min-win coalition, then, for $j \in S^{(\lambda)}$ the domain is $C^{(j)} = [\ell(S^{(\lambda)})+1, n]$. The steps in $S^{(\lambda)}$ are exactly the *inevitable* (or *veto-*) players. If all players in $S^{(\lambda)}$ are steps, then v is the unanimous game of $S^{(\lambda)}$ (with minimal representation $(\bar{m}; \bar{\lambda}) = (1, \ldots, 1, 0, \ldots, 0; \lambda)$.

4. The sums in $S^{(\lambda)}$ determine the nature of the smaller players: the *satellite game* of a sum $j \in S^{(\lambda)}$ is

$$(C^{(j)}, P(C^{(j)}), v^{(j)})$$

where $v^{(j)}$ is represented by $(m_{|C^{(j)}}; m_j)$.

For every sum $j \in S^{(\lambda)}$, $v^{(j)}$ is a homogeneous game (The Basic Lemma). $i > \ell(S^{(\lambda)})$ is a sum w.r.t. v if i is a sum w.r.t. at least one $v^{(j)}$; i is a dummy if i is a dummy in every $v^{(j)}$, and i is a step in any other case.

5. In every homogeneous representation $(m; \lambda)$ of v, sums of the same type have the same weight. Steps of the same type may have different weights, but then they appear or do not appear simultaneously ("as a block") in every min-win coalition.

6. The procedure described in 3. suggests that sums may be replaced in the lex-max min-win coalition $S^{(\lambda)}$ if the players following $\ell^{(\lambda)} = \ell(S^{(\lambda)})$ can muster enough weight. This procedure may be generalized as follows. Let $(m; \lambda)$ be the minimal representation of a homogeneous game $v = v_\lambda^m$. Let $S \in W^m$ and let $\ell = \ell(S)$ again denote the last player in S. Suppose $j \in S$ is such that

$$[j, \ell] \subseteq S, S - j + [\ell + 1, n] \in W. \tag{1.13}$$

Then j is *expendable*; we may replace him in S by an interval of smaller players, thus generating a coalition

$$\rho_j(S) := S - j + [\ell(S) + 1, t] \tag{1.14}$$

where t is uniquely defined by $m([\ell(S) + 1, t]) = m_j$. This procedure is based on the Basic Lemma (Rosenmüller [16]), see Sudhölter [19].

On the other hand, let $T \in W^m$ and suppose that $r \notin T$ satisfies

$$[r + 1, \ell(T)] \subset T. \tag{1.15}$$

Then r is the last *dropout* (denoted by $r = r(T)$) and there is a unique $t' \in [r+1, \ell(T)]$ such that

$$\varphi(T) := T + r - [t', \ell(T)] \tag{1.16}$$

is a min-win coalition. That is, φ inserts the last dropout and cuts off an appropriate tail of T as to generate a min-win coalition. Thus, ρ_j renders j to be the last dropout if he is expendable in S.

Clearly, if r is the last dropout in T, then (he is expendable in $\varphi(T)$ and)

$$\rho_r(\varphi(T)) = T. \tag{1.17}$$

Similarly, if j is expendable in S then (he is the last dropout in $\rho_j(S)$ and)

$$\varphi(\rho_j(S)) = S \tag{1.18}$$

holds true.

We provide a few examples of homogeneous games which may be useful for the purpose of illustrating the general theory developed in the following sections.

Example 1.2. For simpler reading we omit the brackets in a representation $(m; \lambda)$ and indicate the semicolon by a $*$. The characters are indicated by s for sum, τ for step, and d for dummy. For $n = 11$

$$\begin{array}{cccccccccccc} 22 & 11 & 11 & 9 & 2 & 2 & 2 & 2 & 2 & 1 & 1 & {}^*33 \\ s & s & s & s & s & s & s & s & s & s & \tau & \end{array}$$

is a constant sum game—hence a game without steps by convention, since technically the smallest player is the only step. By omitting this smallest player, we obtain for $n = 10$:

$$\begin{array}{ccccccccccc} 22 & 11 & 11 & 9 & 2 & 2 & 2 & 2 & 2 & 1 & {}^*33 \\ s & s & s & s & s & s & s & s & \tau & \tau & \end{array}$$

here the last two players are steps, but of different type. Note that, on the other hand, of course all players with weight 2 are of the same type.

Next, for n = 8, consider

$$\begin{array}{ccccccccc} 12 & 10 & 5 & 3 & 2 & 2 & 1 & 1 & {}^*22 \\ s & s & \tau & s & s & s & s & \tau & \end{array}$$

were a step with weight 5 follows immediately behind the lex-max min-win coalition $S^{(\lambda)}$—this example is treated in more detail in Example 3.7.

For $n = 6$ note the two very similar looking representations

$$\begin{array}{ccccccc} 5 & 4 & 3 & 2 & 1 & 1 & {}^*12 \\ \tau & s & s & \tau & s & \tau & \end{array}$$

and

$$\begin{array}{ccccccc} 5 & 3 & 3 & 2 & 1 & 1 & {}^*11 \\ \scriptstyle T & \scriptstyle S & \scriptstyle S & \scriptstyle T & \scriptstyle S & \scriptstyle T & \end{array}$$

which are nevertheless different—in both cases there is an inevitable (or veto-) player and further steps present.

So far all the above examples are given by their minimal representation. For $n = 15$

$$\begin{array}{cccccccccccccccc} 252 & 252 & 56 & 56 & 56 & 56 & 28 & 28 & 9 & 9 & 5 & 5 & 5 & 2 & 2 & {}^*672 \\ \scriptstyle T & \scriptstyle T & \scriptstyle S & \scriptstyle S & \scriptstyle S & \scriptstyle S & \scriptstyle S & \scriptstyle S & \scriptstyle S & \scriptstyle S & \scriptstyle S & \scriptstyle S & \scriptstyle T & \scriptstyle T & \scriptstyle T & \end{array}$$

is a homogeneous representation but not the minimal one, which is

$$\begin{array}{cccccccccccccccc} 86 & 86 & 32 & 32 & 32 & 32 & 16 & 16 & 5 & 5 & 3 & 3 & 3 & 1 & 1 & {}^*268 \\ \scriptstyle T & \scriptstyle T & \scriptstyle S & \scriptstyle S & \scriptstyle S & \scriptstyle S & \scriptstyle S & \scriptstyle S & \scriptstyle S & \scriptstyle S & \scriptstyle S & \scriptstyle S & \scriptstyle T & \scriptstyle T & \scriptstyle T & \end{array}$$

Similarly,

$$\begin{array}{ccccccccccccc} 170 & 70 & 70 & 31 & 29 & 10 & 4 & 4 & 2 & 2 & 2 & 1 & {}^*240 \\ \scriptstyle T & \scriptstyle S & \scriptstyle S & \scriptstyle T & \scriptstyle T & \scriptstyle S & \scriptstyle S & \scriptstyle S & \scriptstyle S & \scriptstyle S & \scriptstyle T & d & \end{array}$$

is minimally represented by

$$\begin{array}{ccccccccccccc} 50 & 21 & 21 & 8 & 8 & 5 & 2 & 2 & 1 & 1 & 1 & 0 & {}^*71 \\ \scriptstyle T & \scriptstyle S & \scriptstyle S & \scriptstyle T & \scriptstyle T & \scriptstyle S & \scriptstyle S & \scriptstyle S & \scriptstyle S & \scriptstyle S & \scriptstyle T & d & \end{array}$$

—note that the two steps (weight 31 and 29) in the first representation are of equal type, hence get the same weight in the minimal representation.

The largest homogeneous game appearing in reality that we know of is the Thai parliament of 1983, given by

$$92 \quad 73 \quad 56 \quad 24 \quad 18 \quad 15 \quad 4 \quad 3 \quad 2 \quad 1{}^*145$$

which is a non-homogeneous representation of a homogeneous game represented minimally by

$$\begin{array}{ccccccccccc} 23 & 17 & 17 & 6 & 6 & 5 & 1 & 1 & 1 & 1 & {}^*40 \\ \scriptstyle S & \scriptstyle S & \scriptstyle S & \scriptstyle S & \scriptstyle S & \scriptstyle T & \scriptstyle S & \scriptstyle S & \scriptstyle S & \scriptstyle T & \end{array}$$

2. Preliminary results about pre-kernel and kernel

We consider only games without dummies. This section serves to recall some definitions and to prove some elementary properties of the pre-kernel and its relatives.

Definition 2.1. $\mathcal{I}^* = \{x \in \Re^n | x(\Omega) = 1\}$ is the set of *pre-imputations*, and $\mathcal{I} = \{x \in \mathcal{I}^* | x \geq 0\}$ is the set of *pseudo-imputations.*

We write $T_{ij} = \{S \in P | i \in S, j \notin S\}$. Next, for $S \in P$ and $x \in \Re^n$, $e(S,x) = v(S) - x(S)$ is the *excess* of x at S (with respect to some game v).

Definition 2.2. The *maximal* excess of $x \in \Re^n$ with respect to a game v is

$$\mu(x) = \mu(x,v) = \max_{S \in P} e(S,x); \tag{2.1}$$

also, we use for $i, j \in \Omega$, $i \neq j$:

$$s_{ij}(x) = s_{ij}(x,v) = \max\{e(S,x) | S \in T_{ij}\}. \tag{2.2}$$

The corresponding systems of coalitions reaching maximal excess are given by

$$\mathcal{D}(x) = \mathcal{D}(x,v) = \{S \in P | e(S,x) = \mu(x)\} \tag{2.3}$$

and

$$\mathcal{D}_{ij}(x) = \mathcal{D}_{ij}(x,v) = \{S \in T_{ij} | e(S,x) = s_{ij}(x)\}. \tag{2.4}$$

Next, let us recall the definitions for the pre-kernel and its relatives.

1.4

Definition 2.3. Let v be a game.

1. The *pre-kernel* of v is given by

$$\mathcal{P}r\mathcal{K}(v) = \{x \in \mathcal{I}^* | s_{ij}(x) = s_{ji}(x) (i,j \in \Omega, i \neq j)\}. \tag{2.5}$$

2. The *pseudo-kernel* of v is given by

$$\mathcal{P}s\mathcal{K}(v) = \{x \in \mathcal{I} | s_{ij}(x) \leq s_{ji}(x) \text{ or } x_j = 0\, (i,j \in \Omega, i \neq j)\}. \tag{2.6}$$

3. The *kernel* of v is given by

$$\begin{aligned} \mathcal{K}(v) = \{x_j \in \mathcal{I}^* | x_j \geq v_j (j \in \Omega), s_{ij}(x) \leq s_{ji}(x) \text{ or } x_j \\ = v_j (i,j \in \Omega, i \neq j)\}\ . \end{aligned} \tag{2.7}$$

The kernel was introduced in Davis-Maschler [1], see also Maschler-Peleg-Shapley [7], Maschler-Peleg [5], [6] and Peleg [9].

The pseudo-kernel has no satisfying game-theoretical interpretation. We will only use it temporarily to formulate the results of this section—this concept will not be mentioned in the following sections. In fact, the aim of this section is to show that in all cases to be treated essentially all three concepts coincide.

Lemma 2.4. *("Equal treatment–monotonicity") Let v be a directed game and $x \in \mathcal{P}r\mathcal{K}(v)$. Then the following holds true:*

1. *If $i \sim j$, then $x_i = x_j$.*
2. *$x_1 \geq x_2 \geq \cdots \geq x_n$.*

Proof: The pre-kernel respects the desirability relation; see e.g. Peleg [11], Ch.5, Theorem 5.3.5 and Remark 5.3.10. ∎

Lemma 2.5. *Let v be a monotonic directed simple game (without dummies!). If $x \in \mathcal{P}r\mathcal{K}(v)$, then $x \geq 0$.*

Proof:

1st Step:

By Lemma 2.4, we have $x_1 \geq \cdots \geq x_n$. Assuming $x_n < 0$ we would like to end up with a contradiction.

To this end let i_0 be the last player with nonnegative coordinate of x, thus

$$x_{i_0} \geq 0 > x_{i_0+1}. \tag{2.8}$$

Note that player n cannot be a winning player for otherwise (the game is directed!) all players would be winning, hence all would be of the same type—in which case $x_n < 0$ and Lemma 2.4 are not compatible.

Now, fix a coalition $\bar{S}$ such that $n \in \bar{S}$ and

$$e(\bar{S}, x) = \max\{e(S, x) | n \in S\}.$$

The next two steps distinguish two cases according to whether $\bar{S}$ is winning or losing—and in both cases we will eventually end up with a contradiction.

2nd Step:

Assume $\bar{S}$ is winning. Necessarily we have

$$[i_0 + 1, n] \subseteq \bar{S} \tag{2.9}$$

for if some player of the interval is missing, adding him to $\bar{S}$ would increase the excess.

Next, $\bar{S} = \Omega$ cannot occur: Ω has excess 0 and $\{n\}$ has excess $-x_n > 0$ ($\{n\}$ is not winning by the first step!), contradicting the definition of $\bar{S}$. Thus, there is $i \in \Omega - \bar{S}$ and by 2.8 and 2.9:

$$x_i \geq 0. \tag{2.10}$$

We now have $\bar{S} \in T_{ni}$ and $e(\bar{S}, x) = s_{ni}(x)$. Since $x \in \mathcal{P}r\mathcal{K}(v)$, we must be able to find $\hat{S} \in T_{in}$ such that

$$e(\hat{S}, x) = s_{in}(x) = s_{ni}(x) = e(\bar{S}, x) \tag{2.11}$$

holds true.

However, $\hat{S}$ is losing (because n is not a member and adding him increases the excess for any winning coalition), and in addition $(\hat{S} - i) + n$ is *a fortiori* losing (i is stronger than n, as our game is directed). Hence

$$e((\hat{S} - i) + n, x) = e(\hat{S}, x) + x_i - x_n > e(\hat{S}, x) \tag{2.12}$$

since $x_n < 0$ and $x_i \geq 0$ by (2.10). Now, we have the desired contradiction as (2.12) is opposed to the choice of $\hat{S}$, manifested in (2.11). This finishes the 2nd Step.

3rd Step:

Assume now that $\bar{S}$ is losing. Clearly, $\bar{S} \neq \Omega$. But we can state much more, namely

$$[1, i_0] \not\subseteq \bar{S}. \tag{2.13}$$

Indeed, if (2.13) is not true, then pick $i \in \Omega - \bar{S}$. By (2.8) clearly $x_i < 0$ But

$$e((\bar{S} - 1) + i, x) = e(\bar{S}, x) + x_1 - x_i > e(\bar{S}, x) \tag{2.14}$$

again using (2.8) this contradicts the definition of $\bar{S}$. Hence, (2.13) is verified.

Next, pick $j \in [1, i_0] - S$ and let $\tilde{S} \in T_{jn}$ be such that

$$e(\tilde{S}, x) = s_{jn}(x) = s_{nj}(x) = e(\bar{S}, x) \tag{2.15}$$

(this corresponds to (2.11)).

Again, $\tilde{S}$ has to be losing by the same argument as above (n is no member!) and once more the fact that $(\tilde{S} - j) + n$ is losing (j is stronger than n) yields a contradiction, since

$$e((\tilde{S} - j) + n, x) = e(\tilde{S}, x) + x_j - x_n > e(\tilde{S}, x) \tag{2.16}$$

is incompatible with the definition of $\tilde{S}$ in view of (2.15) ∎

Lemma 2.6. *Let v be a monotonic directed simple game (without dummies). If $x \in \mathcal{P}s\mathcal{K}(v)$ and $i \neq j$, then $s_{ij} = s_{ji}$.*

Proof: Let $x \in \mathcal{P}s\mathcal{K}(v)$ and let $i \neq j$. Assume, *per absurdum*, that $s_{ij}(x) > s_{ji}(x)$. Now, as $x \in \mathcal{P}s\mathcal{K}(v)$ we have $x_j = 0$. Choose $\bar{S} \in T_{ij}$ such that

$$e(\bar{S}, x) = s_{ij}(x) \tag{2.17}$$

and $t \in \Omega$ be given by

$$t = \max\{r | x_r > 0\} < n. \tag{2.18}$$

We are going to discuss two cases, both of which end up with a contradiction, proving our Lemma.

1st Case: $[1,t] \not\subseteq \bar{S}$.

In this case, pick $k \in [1,t] - \bar{S}$ and $\tilde{S} \in T_{ki}$ with

$$e(\tilde{S}, x) = s_{ki}(x).$$

Note that $x_k > 0$ implies $s_{ik}(x) \leq s_{ki}(x)$. Thus

$$e(\tilde{S}, x) = s_{ki}(x) \geq s_{ik}(x) \geq e(\bar{S}, x) \tag{2.19}$$

as $\bar{S} \in T_{ik}$.

Now, if $\tilde{S}$ is winning, then so is $\tilde{S} \cup j$, hence

$$e(\tilde{S} \cup j, x) = e(\tilde{S}, x) \geq e(\bar{S}, x) = s_{ij}(x) > s_{ji}(x) \tag{2.20}$$

contradicting the fact that $\tilde{S} \cup j \in T_{ji}$. On the other hand, if $\tilde{S}$ is losing, then $e(\tilde{S}, x) < 0$ as $k \in \tilde{S}$, thus

$$0 > e(\tilde{S}, x) \geq e(\bar{S}, x) = s_{ij}(x) \geq e(\{j\}, x) = -x_j = 0. \tag{2.21}$$

2nd Case: $[1,t] \subseteq \bar{S}$.

In this case $x(\bar{S}) = 1$ and

$$e(\bar{S}, x) = s_{ij}(x) > s_{ji}(x) \geq e(\{j\}, x) \geq 0 = 1 - x(\bar{S}) \geq e(\bar{S}, x) \tag{2.22}$$

which yields a contradiction. ∎

Corollary 2.7. *Let v be a monotonic directed simple game (without dummies). Then*

1. *$\mathcal{P}s\mathcal{K}(v) = \mathcal{P}r\mathcal{K}(v)$.*
2. *If v has no winning players, then*

$$\mathcal{P}s\mathcal{K}(v) = \mathcal{P}r\mathcal{K}(v) = \mathcal{K}(v).$$

3. *If v has no winning players but k inevitable (or veto-) players, then*

$$\mathcal{P}s\mathcal{K}(v) = \mathcal{P}r\mathcal{K}(v) = \mathcal{K}(v) = (\underbrace{\frac{1}{k}, \ldots, \frac{1}{k}}_{k}, 0, \ldots, 0).$$

3. How to avoid winning players

Within this section we shortly deal with the case that winning players are present. *A priori*, this is not excluded for a homogeneous game—we may have several players i with weight $m_i = \lambda$. However, as we are going to show, the pre-kernel of such a game and the pre-kernel of a game obtained by "omitting" the winning players in a suitable way are closely related. Of course, the advantage of the second game is that it is trivially "weakly superadditive" and hence the kernel and the pre-kernel coincide.

Let v be a monotonic directed simple game and let

$$\kappa = \max\{i \in \Omega | \{i\} \in W^m\} \tag{3.1}$$

be the smallest winning player ($\max \emptyset = 0$). Consider the decomposition of Ω into the winning players and the rest

$$\Omega = [1, \kappa] + [\kappa + 1, n] =: \Omega_0 + \Omega_1 \tag{3.2}$$

and the corresponding restrictions

$$v^0 = v|_{\Omega_0} \text{ and } v^1 = v|_{\Omega_1}. \tag{3.3}$$

It is easily verified that $v(\cdot) = (\max v^0(\cdot), v^1(\cdot))$.

In order to exhibit the relationship between the pre-kernels of v, v^0 and v^1, some trivial cases may be excluded.

E.g., if Ω_1 is losing (w.r. to v), then all members of Ω_1 are dummies, v^1 is identically 0 and v^0 equals v (up to omitting the dummies—that is, depending on which version of the restriction is favored).

Thus, the pre-kernel of v is $(\frac{1}{\kappa}, \ldots, \frac{1}{\kappa}, 0, \ldots, 0)$. Similar observations are due if $\Omega_0 = \emptyset$ or $\Omega_1 = \emptyset$.

Theorem 3.1. *Let v be a monotonic directed simple game (on $\Omega = [1, n]$). Let κ denote the smallest winning player and let v^0 and v^1 be defined by (3.2) and (3.3). Assume $\Omega_0 \neq \emptyset \neq \Omega_1$. If $\bar{x} \in \mathcal{P}r\mathcal{K}(v^1)$ and*

$$\bar{\alpha} := \min\{\bar{x}(S) | S \in W(v^1)\} \tag{3.4}$$

then

$$\bar{z} := \frac{1}{\kappa\bar{\alpha} + 1}(\bar{\alpha}, \ldots, \bar{\alpha}, \bar{x}_{\kappa+1}, \ldots, \bar{x}_n) \tag{3.5}$$

is an element of $\mathcal{P}r\mathcal{K}(v)$.

Proof: We have to show that $s_{ij}(\bar{z}) = s_{ji}(\bar{z})$ for all $i, j \in \Omega$. Naturally, this goes by distinguishing all cases according to whether i and j are winning players or not.

1st Case:

If i, $j \in \Omega_0$, then $\bar{z}_i = \bar{z}_j$, hence $s_{ij}(\bar{z}) = s_{ji}(\bar{z})$ is trivial.

2nd Case:

Let $i \in \Omega_0$ and $j \in \Omega_1$. First we have

$$s_{ij}(\bar{z}) = 1 - \bar{z}_i = 1 - \frac{\bar{\alpha}}{k\bar{\alpha}+1}. \tag{3.6}$$

On the other hand, as $\bar{x} \in \mathcal{P}r\mathcal{K}(v^1)$, there is $S \in W(v^1)$ such that $j \in S$ and $\bar{x}(S) = \bar{\alpha}$. Clearly

$$s_{ji}(\bar{z}) = 1 - \frac{\bar{x}(S)}{k\bar{\alpha}+1} = 1 - \frac{\bar{\alpha}}{k\bar{\alpha}+1} = s_{ij}(\bar{z}). \tag{3.7}$$

3rd Case:

Now consider the case that i, $j \in \Omega_1$.

Now, if v^1 has no inevitable (or "veto"-) players, then $s_{ij}(\bar{z}) = e(\bar{S}, \bar{z})$ and $s_{ji}(\bar{z}) = e(\bar{T}, \bar{z})$ for some $\bar{S}$, $\bar{T} \in W(v^1)$. Hence $s_{ij}(\bar{z}) = s_{ji}(\bar{z})$.

If v^1 has inevitable players, then the same argument holds true if i and j both are not inevitable.

If both are inevitable, then $\bar{x}_i = \bar{x}_j$ (Lemma 2.4) and hence $\bar{z}_i = \bar{z}_j$ and $s_{ij}(\bar{z}) = s_{ji}(\bar{z})$.

Finally, if i is inevitable and j is not, then $\bar{x}_j = 0 = \bar{z}_j$. In this case actually $s_{ij}(\bar{z}) = 1 - \frac{\bar{\alpha}}{\kappa\bar{\alpha}+1} = s_{ji}(\bar{z})$. ∎

Theorem 3.2. *Let v be a monotonic directed simple game with smallest winning player κ and v^0 and v^1 be defined by (3.2) and (3.3). Assume $\Omega_0 \neq \emptyset \neq \Omega_1$. Let $\bar{z} \in \mathcal{P}r\mathcal{K}(v)$. Then $\bar{z}(\Omega_1) > 0$ and*

$$\bar{x} := \frac{\bar{z}|\Omega_1}{\bar{z}(\Omega_1)} \in \mathcal{P}r\mathcal{K}(v^1) \tag{3.8}$$

Proof: By the absence of dummies Ω_1 contains a winning coalition, thus $\bar{z}(\Omega_1) > 0$.

Let $\bar{\mathcal{D}} = \{S \subseteq \Omega_1 | e(S, \bar{z}) \geq e(T, \bar{z})(T \subseteq \Omega_1)\}$ and $\bar{s} = e(S, \bar{z})$ for $S \in \bar{\mathcal{D}}$. Let $i \in S \in \bar{\mathcal{D}}$ be such that $\bar{z}_i > 0$. Then

$$\bar{s} = 1 - \bar{z}_1 \tag{3.9}$$

since $s_{1i}(\bar{z}) = s_{i1}(\bar{z})$. We distinguish the following cases:

(i) v^1 has no inevitable(or veto-) player.
In this case $\cap\{S|S \in \bar{\mathcal{D}}\} = \emptyset$ since there cannot be a coalition $S \in \bar{\mathcal{D}}$ with $S \supseteq \{i \in \Omega_1 | \bar{z}_i > 0\}$. Hence, if $i, j \in \Omega_1$, $i \neq j$, then both $s_{ij}(\bar{z})$ and $s_{ji}(\bar{z})$ are attained by winning coalitions w.r.t. v^1, thus $s_{ij}(\bar{x}) = s_{ji}(\bar{x})$.

(ii) v^1 has veto players.
In this case $\cap\{S|S \in \bar{\mathcal{D}}\} \subseteq \{i \in \Omega_1 | i$ is inevitable w.r.t. $v^1\}$, since there is $S \in \bar{\mathcal{D}}$ with $\emptyset \neq S \neq \Omega_1$ as long as not all players of v^1 are inevitable. Consequently $\bar{z}_i = 0$ if $i \in \Omega_1$ is not inevitable and $\bar{z}_j = \bar{z}_k$ for any inevitable players j, k; thus $\bar{x}$ is the unique member of $\mathcal{P}r\mathcal{K}(v^1)$. ∎

Remark 3.3. For the discussion in the following sections we will now always assume that no winning players are present.

4. The kernel for games with steps

This section is devoted to the task of reducing the (pre-) kernel of a homogeneous game with steps to the one of a "smaller" game without steps. More precisely, we show that coordinates of the (pre-) kernel vanish behind the largest step, thus one can restrict oneself to computing the (pre-) kernel of the reduced game obtained by neglecting the players following the first step.

By the results of Section 3, we may restrict ourself to the case that no winning players are present. Also we can clearly dispose of the case in which dummies are present. And finally, in view of Corollary 2.7, we will assume that there are no inevitable players. A further reduction seems useful for notational convenience. Let v be a homogeneous game with steps and let $\underline{\tau} = \underline{\tau}(v)$ be first (largest step). Next, let τ be the smallest player of $\underline{\tau}$'s type. Then $[\underline{\tau}, \tau]$ appear in every min-win coalition either simultaneously or not at all—they form a "block". Now, we shall assume that $\underline{\tau} = \tau$, i.e., the first type of steps consist of one player only. Indeed, the following proofs have to be altered only in an obvious way in order to be carried over to the case of an existing "block". To simplify matters, we begin with

Definition 4.1.

1. A *standard step game* is a homogeneous game with steps, having no dummies, no winning and no inevitable players, such that the first step is the only step of his type.

2. If v is a standard step game then $\tau = \tau(v)$ denotes the first step.

3. If v is a standard game, $\bar{x} \in \mathcal{K}(v)$ and $\bar{x}_\tau > 0$, then $(v, \bar{x})$ will be called a "standard situation".

4. If $(v, \bar{x})$ is a standard situation then $t = t(v) = \max\{i | \bar{x}_i > 0\}$ denotes the smallest player with positive coordinate at $\bar{x}$.

Of course, for any standard step game the kernel and the pre-kernel coincide, so we will mention the kernel only. Clearly, if v is a standard step game, $\bar{x} \in \mathcal{K}(v)$ and $x_\tau = 0$, then $x_i = 0$ for $i \geq \tau$ by Lemma 2.4.

Thus our aim is to show that for any standard situation $(v, \bar{x})$ it follows that $\bar{x}_i = 0 (i > \tau)$.

Let us start out with some preparations.

Definition 4.2. Let $(v, \bar{x})$ be a standard situation. Let $\tau = \tau(v)$ and $t = t(v)$. Define

$$\begin{aligned} \mathcal{M} &= \mathcal{M}(v, \bar{x}) = W^m(v) \cap \mathcal{D}(\bar{x}), \\ \mathcal{M}^+ &= \{S \in \mathcal{M} | S \supseteq [\tau, t]\}, \\ \mathcal{M}^- &= \{S \in \mathcal{M} | S \cap [\tau, t] = \emptyset\}. \end{aligned} \tag{4.1}$$

If we refer to a standard situation, then we will sometimes omit the argument $\bar{x}$, then writing $\mathcal{D}$, $\mathcal{M}$, $\mathcal{M}^+$ etc.

Remark 4.3. Let $(v, \bar{x})$ be a standard situation. If $S \in \mathcal{D}$, then there is $i \in S$ such that $S \cap [1, i] \in \mathcal{M}$. I.e., homogeneity ensures that dropping the smallest players (of largest index) results in a min-win coalition and, as this procedure can but increase the excess, it will result in an $\mathcal{M}$-coalition. The principle of "dropping the smallest players in a winning coalition yields to hitting λ" is part of the Basic Lemma (see [15]), thus we shall refer to it as to the Basic Principle.

Theorem 4.4. *Let $(v, \bar{x})$ be a standard situation. Then $\mathcal{M} = \mathcal{M}^+ + \mathcal{M}^-$ and $\mathcal{M}^+ \neq \emptyset \neq \mathcal{M}^-$.*

Proof: First of all, let us show that $S \in \mathcal{M}$ cannot cut properly into $[\tau, t]$. Indeed, let $S \cap [\tau, t] \neq \emptyset$. As $S \in W^m$ and "steps rule their followers" (Sec.1), we have $\tau \in S$.

If, for some $i \in [\tau + 1, t]$, $i \notin S$, then $S \in \mathcal{M} \cap T_{\tau i}$ and (as $\bar{x}_i > 0$) there is $S' \in \mathcal{D} \cap T_{i\tau}$, clearly $S' \in W$. Now, S' contains some $S'' \in W^m$ and, if $i \notin S''$, then $e(S'') > e(S')$, which is impossible. Thus, $S'' \in W^m \cap T_{i\tau}$—contradicting the fact that τ is a step and "rules his followers".

Thus, $S \in \mathcal{M}$ will either contain $[\tau, t]$ or be disjoint to this interval.

Next, since no veto players are present (and $\bar{x} \geq 0$), it is clear that

$$\bigcap_{S \in \mathcal{M}} S = \emptyset, \tag{4.2}$$

thus $\mathcal{M}^- \neq \emptyset$. On the other hand, any $i \in S \in \mathcal{M}^-$ yields $S \in \mathcal{D} \cap T_{i\tau}$. A min-win coalition $T^0 \in T_{\tau i}$ yields $T^0 \in \mathcal{M} \cap T_{\tau i}$, since $\bar{x}_\tau > 0$ and $\tau \in T^0$, thus $T^0 \in \mathcal{M}^+$. ∎

Lemma 4.5. *There exists $S \in \mathcal{M}^+$ such that $[\tau, \ell(S)] \subseteq S$.*

Proof: Pick $S \in \mathcal{M}^+$ and let $r = r(S)$ be the last dropout of S. If $r \in [\tau, \ell(S)]$, then $r \notin [\tau, t]$ (since $S \supseteq [\tau, t]$), thus $\bar{x}_r = 0$ and $\bar{x}_i = 0$ for all i with $r < i$.

Applying the operation φ to S which inserts r and omits smaller players (cf. Sec.1) does, therefore, not change the excess. Also, $\varphi(S) \in \mathcal{M}^+$. We may continue this procedure, until no dropout behind τ is missing in S. ∎

Remark 4.6. In particular, the shortest coalition (the one of minimal length) in $\mathcal{M}^+$ satisfies the condition specified by Lemma 4.5. Clearly, the procedure applied in 4.5 will work in a different context. I.e., if we take $S \in \mathcal{M}^+$ with certain additional properties, then we may frequently assume or conclude that $[\tau, \ell(S)] \subseteq S$, using φ to replace dropouts if necessary. This will be called the Full Tail Principle.

Now, in order to begin the discussion, it should be noted that we have already treated one important, though not very enlightening case. This is the one in which inevitable players are present, and of course Corollary 2.7 shows that elements of the kernel have zero coordinates for all players following τ.

For, indeed, in a standard step game with inevitable players, τ is an element of $S^\lambda = [1, \ell^\lambda]$, τ is the only step in S^λ and the only inevitable player (of course $\tau = 1$ in this case).

For didactical reasons we shall now treat the case that τ follows immediately behind the lex-max min-win coalition $S^{(\lambda)}$, i.e., $\tau = \ell^{(\lambda)} + 1$. Although this case is subsumed under the general proof provided for Theorem 4.14, studying this particular situation is very enlightening. The proof of our main result is much simpler if $\tau = \ell^{(\lambda)} + 1$ and still exhibits some structure of the general situation. We believe that it will help the reader to follow the main exposition and hence we shall treat it in advance.

Example 4.7. Let $(m;\lambda) = (12, \quad 10, \quad 5, \quad 3, \quad 2, \quad 2, \quad 1, \quad 1; \quad 22)$.

We identify coalitions with $0-1$-vectors ("characteristic functions of coalitions") thus clearly

$$S^{(\lambda)} = \begin{matrix} 1 & 1 & 0 & 0 & 0 & 0 & 0 & 0 \\ 12 & 10 & 0 & 0 & 0 & 0 & 0 & 0 \end{matrix}$$

and since both of them can be replaced, players 1 and 2 are sums. However, the coalition of minimal length containing player 3 (weight 5) is

$$\begin{matrix} 1 & 0 & 1 & 1 & 1 & 0 & 0 & 0 \\ 12 & 0 & 5 & 3 & 2 & 0 & 0 & 0 \end{matrix}$$

and there is not enough weight among the smaller players in order to replace player 3—thus player 3 is a step, that is $\tau = \tau(v) = 3 = 2 + 1 = \ell^{(\lambda)} + 1$.

The reader may want to follow the proof of Lemma 3.8 while viewing the above example.

Lemma 4.8. *Let $(v, \bar{x})$ be a standard situation. Assume*

$$\tau = \ell^{(\lambda)} + 1 \tag{4.3}$$

Then $\bar{x}_i = 0$ for all $i > \tau$.

Proof:

1st Step:

In this particular case $S^{(\lambda)}$ is the only min-win coalition not containing τ, thus

$$\mathcal{M}^- = \{S^{(\lambda)}\}. \tag{4.4}$$

2nd Step: The shortest min-win coalition containing τ is

$$S^\tau := (S^{(\lambda)} - \ell^{(\lambda)}) + [\tau, s] \tag{4.5}$$

with suitable $s > \tau$. All elements in $[\tau + 1, s]$ are sums, for otherwise they would be of the same type as τ. Since S^τ is shortest containing them, they may be replaced by players following $s-$ without disturbing the min-win property, that is.

3rd Step:

By (4.4), $S^{(\lambda)} \in T_{\ell^{(\lambda)},\tau} \cap \mathcal{D}$, consequently, there is $\bar{S} \in T_{\tau,\ell^{(\lambda)}} \cap \mathcal{D}$.

By the Basic Principle, $\bar{S} \in \mathcal{M}$, thus $\bar{S} \in \mathcal{M}^+$. By the Full Tail Principle $\bar{S} \supseteq [\tau, \ell(\bar{S})]$. Now, $\ell^{(\lambda)}$ is the only player from $S^{(\lambda)}$ missing in

$\bar{S}$ (since τ cannot be replaced by players following s, no player $i \in [1, \tau-1]$ can be replaced by players following s)—thus $\bar{S}$ has the form

$$\bar{S} = (S^{(\lambda)} - \ell^{(\lambda)}) + [\tau, \ell(S)]$$

and (since both are min-win), $\bar{S} = S^\tau$.

In view of the 2nd Step any $i \in [\tau+1, \ell(\bar{S})]$ can be replaced by players to the right of $\ell(\bar{S})$—but since $t \in \bar{S}$, these have zero $\bar{x}$-coordinates. Hence $\bar{x}_i = 0$ (otherwise the excess increases). Thus $t = \tau$. ∎

The general case, to be tackled by a series of auxiliary statements, is of course more involved—but some flavor of the simple proof offered in Lemma 4.8 is always present.

For instance, note that $S^\tau = \bar{S}$ as constructed in the third step has the largest dropout possible for $S \in \mathcal{M}^+$ and S obeying the Full Tail Principle. As a result $\varphi(\bar{S}) = S^{(\lambda)}$.

Our first aim is to imitate this idea on a more general basis. This is attempted by the following Definition 4.9 and Theorem 4.10.

Definition 4.9. Let $(v, \bar{x})$ be a standard situation. Define

$$\mathcal{M}^{++} := \{S \in \mathcal{M}^+ | [\tau, \ell(S)] \subseteq S\}, \tag{4.6}$$

$$\bar{r} := max\,\{r(S) | S \in \mathcal{M}^{++}\}. \tag{4.7}$$

Note that $\mathcal{M}^{++} \neq \emptyset$ by Lemma 4.5!

Theorem 4.10. *Let $\bar{S} \in \mathcal{M}^{++}$ be such that $r(\bar{S}) = \bar{r}$. Then $\varphi(\bar{S}) \in \mathcal{M}^-$ and $\ell(\varphi(\bar{S})) = \bar{r}$.*

As a consequence, we have

$$m_{\bar{r}} = m([\bar{r}+1, \ell(\bar{S})]). \tag{4.8}$$

Proof:

1st Step:

Since $\bar{S} \in T_{\tau,\bar{r}}$, there is $\bar{T} \in T_{\bar{r},\tau}$ such that $\bar{T} \in \mathcal{M}$: use the Basic Principle. In view of Theorem 4.4,

$$\bar{T} \in \mathcal{M}^-. \tag{4.9}$$

2nd Step:

We want to show that

$$\ell(\bar{T}) = \bar{r} \tag{4.10}$$

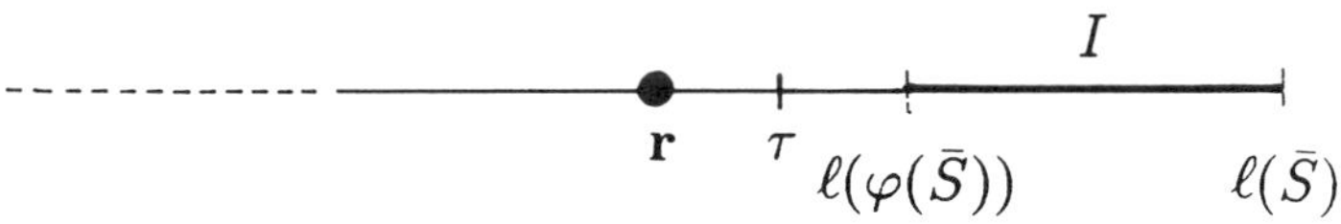

Fig. 1 Coalition $\bar{S}$

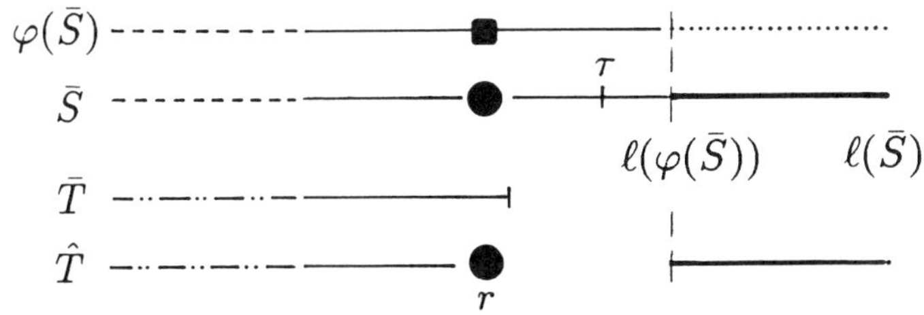

Fig. 2 $\bar{S}, \bar{T}$ and their derivates

holds true. To this end, we are going to show that $[\bar{r}+1, \tau-1] \cap \bar{T} = \emptyset$. Assume on the contrary that $i \in [\bar{r}+1, \tau-1] \cap \bar{T}$, then i separates τ via $\bar{T}$. Hence, there exists $\tilde{T} \in T_{\tau i} \cap \mathcal{D}$. By the Basic Principle, we may assume that $\tilde{T} \in \mathcal{M}$ holds true. But then the fact that $\bar{r} < i \leq r(\tilde{T})$ contradicts the maximality of $\bar{r}$ (i.e. (4.7)). Thus, we have indeed verified (4.10).

3rd Step:

Define now

$$\bar{I} := [\ell(\varphi(\bar{S})) + 1, \ell(\bar{S})]. \tag{4.11}$$

This interval serves to replace $\bar{r}$ in $\bar{S}$, thus

$$m_{\bar{r}} = m(\bar{I}). \tag{4.12}$$

Therefore, if we put

$$\hat{T} := (\bar{T} - \bar{r}) + \bar{I},$$

then clearly $\hat{T} \in W^m$ in view of (4.10) and (4.12).

4th Step:

Next, it is our aim to show that $\hat{T}$ as well as $\varphi(\bar{S})$ have maximal excess.

Indeed, because $\bar{S}, \bar{T} \in \mathcal{M}$, we have

$$\begin{aligned} e(\bar{S}) \geq e(\varphi(\bar{S})) &= e(\bar{S}) - \bar{x}_{\bar{r}} + x(\bar{I}), \\ e(\bar{T}) \geq e(\hat{T}) &= e(\bar{T}) + \bar{x}_{\bar{r}} - x(\bar{I}). \end{aligned} \tag{4.13}$$

It follows that necessarily all inequalities in (4.13) must be equations, that is

$$\varphi(\bar{S}), \hat{T} \in \mathcal{M} \tag{4.14}$$

(and $\bar{x}_{\bar{r}} = x(\bar{I})$).

5th Step:

Now $\hat{T}$

- has maximal excess,
- contains $\ell(\bar{S})$, hence τ;
- is a member of $\mathcal{M}^+$,
- contains $[\tau, \ell(\hat{T})]$,
- hence is in $\mathcal{M}^{++}$.

Therefore, $\hat{T}$ cannot have dropouts in $[\bar{r}+1, \tau-1]$, for this would contradict the maximality of $\bar{r}$, i.e., (4.7).

We conclude that necessarily

$$\bar{I} = [\bar{r}+1, \ell(\bar{S})] \tag{4.15}$$

must be satisfied. In view of (4.11), this means $\ell(\varphi(\bar{S})) = \bar{r}$; this is what we wanted to prove. Clearly, $\tau \notin \varphi(\bar{S})$ and thus $\varphi(\bar{S}) \in \mathcal{M}^-$ (by Theorem 4.4). ∎

In order to proceed with our exposition we will now draw on the theory of homogeneous games. The simple shape of the situation as presented in Lemma 4.8 is no longer prevailing, nevertheless our development so far exhibits "coalition $\bar{S}$"—which also appears in 4.8. What is the natural generalization of S^τ as defined in (4.5)?

Theorem 4.11. *Let v be a standard step game. There exists a sequence of min-win coalitions*

$$S_1, \ldots, S_\tau \in W^m$$

with the following properties:

1. *$S_1 = S^\lambda$.*
2. *For every $k \in [1, \ldots, \tau]$, the system*

   $$\mathcal{S}_k := \{S_i | i \in [1, k], k \in S_i\}$$

   is nonempty.
3. *For all $k \in [1, \ldots, \tau - 1]$, among all $S_i \in \mathcal{S}_k$ with minimal length, let Si_0 be the one with minimal index, then*

   $$S_{k+1} = \rho_k(S_{i_0}).$$

Proof: This follows from Sudhölter [19], Theorem 2.3, Definition 2.4; see also the formulation in [18] ∎

Definition 4.12.

1. Among all coalitions of S_τ with minimal length, let S^τ be the one which has minimal index, i.e., in particular

$$\ell(S^\tau) \geq \tau. \tag{4.16}$$

2. Let $r_a < \cdots < r_1$ denote the dropouts of S^τ (the last dropout is enumerated first since $\varphi = \varphi^1$ will reestablish him; thus $r_1 = r(S^\tau)$). Write $0 =: r_{a+1}$, $r_0 := \tau$ for convenience.

Remark 4.13.

1. If $\varphi^k(S^\tau)$ denotes the k'th iterate of φ applied to S^τ, $\varphi^k(S^\tau) \ni r_k$ and $\varphi^k(S^\tau)$ has smallest length among *all* min-win coalitions containing r_k. Note that $\varphi^a(S^\tau)$ is a min-win coalition without dropouts. (See Sudholter [19]).

2. If $\ell(S^\tau) > \tau$, then $\tau + 1$ is expendable in S^τ. For, S^τ is then certainly a shortest min-win coalition containing $\tau + 1$. If $\tau + 1$ were a step then τ and $\tau + 1$ would never be separated ("steps rule their followers"), hence—as $\tau + 1$ differs in type from τ—$\tau + 1$ has to be sum. As such he has to be expendable in the shortest coalition containing him.

3. Consider a standard situation $(v, \bar{x})$. Suppose that we can find $\hat{S} \in \mathcal{M}^+$ such that $\ell(\hat{S}) = \ell(S^\tau)$. We know that $t = t(v) \in \hat{S}$ and—thus either $t = \tau$ or $t > \tau$, $\tau + 1 \in \hat{S}$, and $\tau + 1$ is expendable in $\hat{S}$ by the argument presented above in 2. Replacing $\tau + 1$ by players behind t will, however, increase the excess properly—thus we have necessarily $\tau = t$.

That is: if there is $\hat{S} \in \mathcal{M}^+$ with $\ell(\hat{S}) = \ell(S^\tau)$, then $t = \tau$!

This reasoning again is quite analogous to the one offered in the 3rd Step of the proof of Lemma 4.8.

Theorem 4.14. *Let $(v, \bar{x})$ be a standard situation. Then $t(v) = \tau(v)$. That is, for any standard step game, the coordinates of kernel payoffs vanish behind the first step.*

Proof: Let S^τ and $0 = r_{a+1} < r_a < \cdots < r_1 < r_0 = \tau$ be defined by Definition 4.12; thus in particular S^τ is a shortest coalition containing τ and has smallest index in the family defined by Theorem 4.11 among all coalitions with this property.

Next, choose $\bar{S} \in \mathcal{M}^{++}$ such that $r(\bar{S}) = \bar{r}$ (cf. (4.6), (4.7)), thus by Theorem 4.10 we know that $\varphi(\bar{S}) \in \mathcal{M}^-$ and $\ell(\varphi(\bar{S})) = \bar{r}$.

Now, fix $k \in [0, a]$ such that

$$r_{k+1} < \bar{r} \le r_k. \tag{4.17}$$

Accordingly, we are going to distinguish several cases.

1st Case:

If $k = a$, i.e., $\bar{r} \le r_a$, then we argue as follows:

In this case in view of $\varphi^a(S^\tau) = S^\lambda$ (Remark (4.13)) and $\ell(\varphi(\bar{S})) = \bar{r}$ (Theorem (4.10)), it follows that $\varphi(\bar{S}) = S^\lambda$.

Then clearly $\bar{S} = S^\tau$; and Remark 4.13 shows that $t = \tau$.

So here, the argument is quite direct and the attentive reader will have observed that Lemma 4.8 actually deals with this case.

In the following we may now assume that $k < a$.

2nd Case:

Consider now the situation in which

$$r_{k+1} \notin \bar{S} \tag{4.18}$$

holds true. Observe that $r_{k+1} \notin \varphi(\bar{S})$, since $r_{k+1} < \bar{r} \notin \bar{S}$; moreover

$$\ell(\varphi^k(S^\tau)) \ge r_k \ge \bar{r} = \ell(\varphi(\bar{S})).$$

Thus we may draw the following sketch.

Now, take $\varphi(\bar{S})$ and $\varphi^k(S^\tau)$ into consideration. Both do not contain r_{k+1}. If we apply φ to $\varphi^k(S^\tau)$, then we obtain a shortest coalition such that r_{k+1} is expendable (Theorem (4.11)). Hence, $\varphi(\bar{S})$ cannot be shorter than $\varphi^k(S^\tau)$, that is

$$\ell(\varphi^k(S^\tau)) = \bar{r}. \tag{4.19}$$

Since $\bar{S} \ni \tau$, this can only mean that $k = 1$ and $\ell(\bar{S}) = \ell(S^\tau)$. Again we fall back on Remark 4.13 in order to conclude that $t = \tau$.

3rd Case:

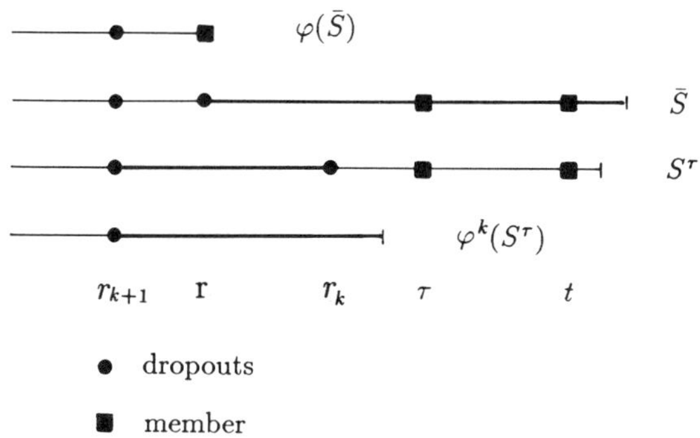

Fig. 3. $\bar{S}$, S^τ and their derivates

Suppose that $r_k = \bar{r}$, then $\ell(\varphi^k(S^\tau)) = \bar{r}$. Again, as $\ell(\varphi(\bar{S})) = \bar{r}$, we must have $k = 1$ and $\ell(\bar{S}) = \ell(S^\tau)$. The argument now proceeds analogously to the 2nd Case.

4th Case:

It remains to consider the case that $r_{k+1} \in \bar{S}$ and $\bar{r} < r_k$. This is the only one that requires a somewhat more elaborate argument.

Now, since $\bar{S} \in T_{r_{k+1}\bar{r}}$, there is $\bar{T} \in \mathcal{M} \cap T_{\bar{r}r_{k+1}}$. Since $\bar{T}$ is minimally winning, this coalition contains a "tail" $[i, \ell(\bar{T})] \cap \bar{T}$ exactly replacing r_{k+1}, thus

$$(\bar{T} + r_{k+1}) \cap [1, i-1] \in W^m(v)$$

and

$$\ell(\bar{T}) \geq \ell(\varphi^k(S^\tau)) \geq r_k \tag{4.20}$$

since $\varphi^k(S^\tau)$ is shortest such that r_k is contained in it (Theorem 4.11).

We claim that $\bar{T} \in \mathcal{M}^+$. Indeed, otherwise (i.e., if $\bar{T} \in \mathcal{M}^-$)$\ell(\bar{T}) < \tau$— since "steps rule their followers"—and thus $s_{\ell(\bar{T})\tau} = s_{\tau\ell(\bar{T})}$, a contradiction to the maximality of $\bar{r}$ (i.e., (4.7)).

Thus $\bar{T} \in \mathcal{M}^+$. By the Full Tail Principle we may immediately assume that

$$\bar{T} \in \mathcal{M}^{++}, \tag{4.21}$$

i.e., $[\tau, \ell(\bar{T})] \subseteq \bar{T}$. That is, $\bar{T}$ has dropouts only to the left of τ.

Next, the definition of $\bar{r}$ together with the fact that $\bar{T}$ is a member of $\mathcal{M}^{++}$ directly implies $r(\bar{T}) < \bar{r}$ and (since $m([\bar{r}+2, n]) < m_{\bar{r}} - m_{\bar{r}+1} + m_\tau \leq m_{\bar{r}} \leq m_{r(\bar{T})}$)

$$\ell(\varphi(\bar{T})) \leq \bar{r} \tag{4.22}$$

Hence $r(\bar{T}) = r_{k+1}$, for otherwise

$$\ell(\varphi(\bar{T})) \geq \ell(\varphi^k(S^\tau)) \geq r_k > \bar{r}.$$

Therefore, we have only two alternatives: either

$$\ell(\varphi(\bar{T})) = \ell(\varphi^{k+1}(S^\tau)),$$

then $k = 0$ and $\ell(\bar{T}) = \ell(S^\tau)$—a situation we know how to deal with via Remark 4.13, or

$$\ell(\varphi(\bar{T})) > \ell(\varphi^{k+1}(S^\tau)).$$

Then we come up with

$$\begin{aligned} m([\bar{r}+1, n]) &\geq m([\bar{r}+1, \ell(\bar{T})] \\ &\geq m(S^\tau \cap [\bar{r}+1, \ell(\bar{T})]) + m_{\ell(\varphi(\bar{T}))} \end{aligned}$$

(by (4.22))

$$\geq m(S^\tau \cap [\bar{r}+1, \ell(\bar{T})]) + m_{\bar{r}}$$

(again by (4.22))

$$\begin{aligned} &\geq m_\tau + m_{\bar{r}} \\ &= m_\tau + m_{\bar{r}+1} + \cdots + m_{\ell(\bar{S})} \end{aligned}$$

(by Theorem 4.10),

i.e.

$$m([\ell(\bar{S})+1, n]) \geq m_\tau. \tag{4.23}$$

But (4.23) contradicts the fact that τ is a step. ∎

5. The Reduction Theorem

Within this section we draw the conclusions and collect results in order to finally formulate Theorem 5.5, the main theorem of this paper.

Definition 5.1. Let v be a standard step game, $\tau = \tau(v)$. Let

$$\begin{aligned} \tilde{m} &:= (m_1, \ldots, m_\tau) = m_{|[1,\tau]}, \\ \tilde{\lambda} &:= \lambda - m[\tau+1, n], \\ \tilde{v}^{(\tau)} &:= v_{\tilde{\lambda}}^{\tilde{m}}. \end{aligned}$$

$\tilde{v}^{(\tau)}$ is the *truncation* of v at τ or for short the *truncated game.*

If $(v, \bar{x})$ is a standard situation, then the truncated game equals the *reduced game* (see Peleg [11], Definition 3.6.8; note that the pre-kernel has the "reduced game property")—more precisely and more generally:

Theorem 5.2. *Let v be a standard step game and $\bar{x} \in \mathcal{K}(v)$. Then*

1. $\bar{x}_i = 0(i > \tau = \tau(v))$,

2. $\bar{x}_{|[1,\tau]} \in \mathcal{K}(\tilde{v}^{(\tau)})$.

This follows immediately from Theorem 4.14.

Remark 5.3. It is important to note that the truncated game $\tilde{v}^{(\tau)}$ is a homogeneous game without dummies; this has been shown in Rosenmüller-Sudhölter [18], Lemma 3.7 (see also Corollary 3.9). In addition $\tilde{v}^{(\tau)}$ is a game without steps. Note that truncation is not always the same as reduction, in particular if we reduce "behind a sum", homogeneity may be destroyed! Note the different version of a truncation presented in [18] for sums.

Naturally, we should set out for the "converse" of Theorem 5.2, i.e., we want to prove

Theorem 5.4. *Let v be a standard step game and $\tilde{v}^{(\tau)}$ the truncated game. If*

$$\tilde{x} = (\tilde{x}_1, \ldots, \tilde{x}_\tau) \in \mathcal{K}(\tilde{v}^{(\tau)}), \text{then } \bar{x} = (\tilde{x}_1, \ldots \tilde{x}_\tau, 0, \ldots, 0) \in \mathcal{K}(v).$$

Proof:

1st Step:

We use ~ to indicate the quantities belonging to $\tilde{v}^{(\tau)}$, e.g.,

$$\tilde{\mathcal{D}} = \{S \in W^m(\tilde{v}^{(\tau)}) = \tilde{W}^m | \tilde{e}(S, \tilde{x}) = \mu(\tilde{x}, \tilde{v}) = \tilde{\mu}\}$$

etc. It is not hard to see that

$$\cap \{S | S \in \tilde{\mathcal{D}}\} = \emptyset, \cap \{S | S \in \mathcal{D}\} = \emptyset \tag{5.1}$$

$$\mu = \tilde{\mu} > 0, \tag{5.2}$$

holds true. Hence, for $i \neq j$, $i \in [\tau + 1, n], j \in [1, n]$

$$s_{ij}(\bar{x}) = \mu = \mu(\bar{x}). \tag{5.3}$$

Using the fact that $\tilde{x} \in \mathcal{K}(\tilde{v}^{(\tau)})$, we immediately have

$$s_{ij}(\bar{x}) = s_{ji}(\bar{x}) \tag{5.4}$$

for $i \neq j$, $i, j \in [1, \tau]$. Therefore, it remains to show that

$$s_{ij}(\bar{x}) = \mu(\bar{x}) \qquad (i \in [1, \tau], j \in [\tau + 1, n]). \tag{5.5}$$

A further reduction is obtained by observing that

$$s_{ik}(\bar{x}) \geq s_{ik-1}(\bar{x}) \qquad (i \in [1,\tau], k \in [\tau+2,n]) \tag{5.6}$$

so that presently all that remains to be shown is that

$$s_{i\tau+1}(\bar{x}) = \mu(\bar{x}) \qquad i \in [1,\tau] \tag{5.7}$$

holds true.

2nd Step:

Next we claim that it suffices to show that

$$s_{\tau\tau+1} = \mu = \mu(\bar{x}) \tag{5.8}$$

holds true.

To this end assume that, for some $i < \tau$ we have $s_{i\tau+1}(\bar{x}) < \mu(\bar{x})$. Then we will immediately show that $s_{\tau\tau+1} < \mu$ is a consequence.

Indeed, $s_{i\tau+1}(\bar{x}) < \mu$ and $s_{\tau\tau+1} = \mu$ is incompatible as follows:

Pick $\bar{S} \in \mathcal{D} \cap T_{\tau\tau+1}$. Clearly, $i \notin \bar{S}$, hence $\tilde{s}_{\tau i}(\tilde{x}) = s_{\tau i}(\bar{x}) = e(\bar{S}) = \mu = \tilde{\mu}$. Since $\tilde{x} \in \mathcal{K}(\tilde{v}(\tau))$, we find $T \in \mathcal{D} \cap T_{i\tau}$. Eliminate players to the right of τ (smaller ones) from T. We obtain a winning coalition, for $\tau \notin T$ and steps rule their followers. Hence, we obtain a coalition, say T', with $T' \in \mathcal{D} \cap T_{i\tau+1}$—contradicting $s_{i\tau+1} < \mu$. This shows that it suffices to prove 5.8.

3rd Step:

There is no problem with the proof of 5.8 if $x_\tau = 0$. For in this case

$$\tilde{s}_{\tau j}(\tilde{x}) = \tilde{\mu} \qquad (j \in [1,\tau-1]) \tag{5.9}$$

and, consequently for all $j \in [1,\tau-1]$ it follows that

$$s_{\tau\tau+1} \geq s_{j\tau+1} \geq s_{j\tau} = \tilde{s}_{j\tau} = \tilde{\mu} = \mu \tag{5.10}$$

holds true.

4th Step:

Hence, we will now set out to prove 5.8 assuming $\tilde{x}_\tau > 0$. Analogously to Definition 4.2, define

$$\mathcal{M}^+ := \{S \in \mathcal{M} | \tau \in S\}, \mathcal{M}^- := \{S \in \mathcal{M} | \tau \notin S\}, \tag{5.11}$$

Such that $\mathcal{M} = \mathcal{M}^+ + \mathcal{M}^-$ and, as is easily verified, $\mathcal{M}^+ \neq \emptyset$ and $\mathcal{M}^- \neq \emptyset$. Similarly, copying 4.9, we put

$$\begin{aligned} \mathcal{M}^{++} &:= \{S \in \mathcal{M}^+ | [\tau, \ell(S)] \subseteq S\} \\ \bar{r} &:= \max\{r(S) | S \in \mathcal{M}^{++}\}. \end{aligned} \tag{5.12}$$

Of course, the Full Tail Principle ensures that $\mathcal{M}^{++} \neq \emptyset$.

Referring to Theorem 4.11 and Remark 4.13, we may now complete the proof by producing some $\hat{S} \in \mathcal{M}^+$ with $\ell(\hat{S}) = \ell(S^\tau)$. For again for such an $\hat{S}$ either $\ell(\hat{S}) = \tau$ or $\tau + 1$ will be expendable without increasing the excess.

The construction can, however, be completed quite analogously to the one presented in the proof of Theorem 4.14. Whenever within the course of this proof reference is made to $\bar{x}$ being an element of $\mathcal{K}(v)$, then, for our present problem the coalition being produced by this argument is already available by $\tilde{x} \in \mathcal{K}(\tilde{v}^{(\tau)})$. (Note that in the course of 4.14 and the previous theorems, essentially only players preceding τ are manipulated in their role as dropouts.) ∎

Theorem 5.5. *Let v be a standard step game. Then*

$$\mathcal{K}(v) = \{(\tilde{x}, 0) | \tilde{x} \in K(\tilde{v}^{(\tau)})\}.$$

6. REFERENCES

[1] M. Davis and M. Maschler, "The kernel of a cooperative game," *Naval Research Logistics Quarterly* **12** (1965) 223–259.

[2] J. R. Isbell, "A class of majority games," *Quarterly Journal Math.* **7** (1956) 183–187.

[3] J. R. Isbell, "A class of simple games," *Duke Math. Journal* **25** (1958) 423–439.

[4] J. R. Isbell, "On the enumeration of majority games," *Math Tables Aids Comput.* **13** (1959) 21–28.

[5] M. Maschler and B. Peleg, "A characterization, existence proof, and dimension bounds for the kernel of a game," *Pacific J. Math.* **18** (1966) 289–328.

[6] M. Maschler and B. Peleg, "The structure of the kernel of a cooperative game," *SIAM J. Appl. Math.* **15** (1967) 569–604.

[7] M. Maschler, B. Peleg, and L. S. Shapley, "Geometric properties of the kernel, nucleolus, and related solution concepts," *Math. of O.R.* **4** (1979) 303–338.

[8] A. Ostmann, "On the minimal representation of homogeneous games," *International J. Game Theory* **16** (1987) 69–81.

[9] B. Peleg, "On the kernel of constant-sum simple games with homogeneous weights," *Ill. Journal Math.* **10** (1966) 39–48.

[10] B. Peleg, "On weights of constant-sum majority games," *SIAM J. Appl. Math.* **16** (1968) 527–532.

[11] B. Peleg, "Introduction to the theory of cooperative games," Research Memoranda No. 81–88, Center for Research in Mathematical Economics and Game Theory, The Hebrew University, Jerusalem, Israel.

[12] B. Peleg and J. Rosenmüller, "The least-core, nucleolus, and kernel of homogeneous weighted majority games," *Games and Economic Behavior* **4** (1992) 588–605.

[13] J. Rosenmüller, "Extreme games and their solutions," *Lecture Notes in Economics and Math. Systems* **145** Springer Verlag, 1977.

[14] J. Rosenmüller, "On homogeneous weights for simple games," Working paper No. 115, Institute of Mathematical Economics, University of Bielefeld, 1982.

[15] J. Rosenmüller, "Weighted majority games and the matrix of homogeneity," *Zeitschrift für Operations Research* **28** (1984) 123–141.

[16] J. Rosenmüller, "Homogeneous games: recursive structure and computation," *Math. of O.R.* **12** (1987) 309–330.

[17] J. Rosenmüller, "Homogeneous games with countably many players," *Math. Social Sciences* **17** (1989) 131–159.

[18] J. Rosenmüller and P. Sudhölter, "The nucleolus of homogeneous games with steps," Working Paper No. 202, Insitute of Mathematical Economics, University of Bielefeld, 1991.

[19] P. Sudhölter, "Homogeneous games as anti step functions," *International J. Game Theory* **18** (1989) 433–469.

[20] J. von Neumann and O. Morgenstern, *Theory of games and economic behavior*, Princeton University Press, 1944.

Essay 14

On the Locus of the Nucleolus

Jos Potters[1] and Stef Tijs[2]

ABSTRACT. In this paper we prove some properties of the locus of the nucleolus of a set C that can be the core of a cooperative game. We show that there are two subsets of C, the kernel and the least core, only dependent on the set C that contain this locus. From this fact we derive that for a set C which is the core of *convex game* the locus consists of one point. We also give a necessary and sufficient condition that the locus consists of one point.

1. Introduction

It was during the Oberwolfach Conference in January 1989 that Michael Maschler forwarded the following question. If the core of a cooperative game is given what can be said about the position of the nucleolus? In this paper we will give a partial answer to that question. We prove that every set C that can be the core of a cooperative game contains two subsets, the kernel and the least core of C, only dependent on C, which contain all nucleoli of games with core C. In special cases, when the intersection of the kernel and the least core consists of one point we can infer that 'the core determines the nucleolus.'

Let N be a finite player set and C be a (nonempty) compact convex set in $\Re^N$. For every coalition $T \subset N$ we define $a_T := \min\{(e_T, x) \mid x \in C\}$, where e_T is the characteristic vector of the coalition T, *i.e.*, $(e_T)_i = 1($ or $0)$ if $i \in T$ (or $i \notin T$). We call the set C *a pre-core* if

$$C = \{x \in \Re^N \mid (e_N, x) = a_N, (e_T, x) \geq a_T \text{ for all } T \subset N\} .$$

It is immediately clear that the game v_C defined by $v_C(S) = a_S$ for all $S \subset N$ has core C and that the game v_C is an exact game (*i.e.*, for every

[1]KUN Department of Mathematics, University of Nijmegen, Toernooiveld, 6525 ED Nijmegen, The Netherlands

[2]KUB Faculty of Economics, University of Tilburg, P. O. Box 90153, 5000 LE Tilburg, The Netherlands

coalition S there is a core element $x \in C$ with $x(S) := \sum_{i \in S} x_i = v_C(S)$ (Schmeidler [10]). As one can see from the definition, the exact game v_C is uniquely determined by the set C and C is conversely the core of v_C:

There is a one-to-one correspondence between pre-cores and exact games.

The key concept of this paper will be the *locus of the nucleolus* in C. If C is a pre-core, the *locus of the nucleolus* $\mathcal{L}\text{oc}\ \mathcal{N}\text{uc}\ (C)$ is the set of points $x \in C$ which can be the nucleolus of a cooperative game with core C.

The following example from Maschler, Peleg, and Shapley [5] shows that $\mathcal{L}oc\mathcal{N}uc(C)$ may consist of more than one point, *i.e.*, the core of a game does not determine the nucleolus of a game.

Example 1.1. Let C be the set $\{(s, 1-s, s, 1-s) \mid \frac{1}{4} \leq s \leq 1\}$. For $0 \leq t \leq \frac{1}{4}$, we define the games v_t by: $v_t(N) = 2$, $v_t(123) = 1+t$, and $v_t(S) = 1$ for $S \in \{(124), (134), (234), (12), (23), (34), (14)\}$. Furthermore, $v_t(13) = \frac{1}{2}$ and $v_t(S) = 0$ for all other coalitions. It is an easy exercise to prove that the core of all the games v_t is C and that the nucleolus of v_t corresponds to $s = \frac{1}{2}(1+t)$. These points are in the set $\mathcal{L}oc\mathcal{N}uc(C)$. The reason that the nucleolus "moves" when t increases is that coalition $S = (123)$ is redundant for the determination of the core (as long as $t \leq \frac{1}{4}$) but is influencing the position of the nucleolus. We conclude that *the locus of the nucleolus of a pre-core C may consist of more than one point.* Furthermore, one can compute the $\mathcal{L}oc\mathcal{N}uc(C)$ to consist of the points $s \in [\frac{1}{2}, \frac{3}{4}]$.

In the next section we will introduce two subsets of C, the *kernel* and the *least core* of C which satisfy the properties:

(1) These sets contain the locus of the nucleolus $\mathcal{L}oc\mathcal{N}uc(C)$.

(2) They are only dependent on the pre-core C.

2. The kernel and the least core of a pre-core C

The first proposition is well known (cf. Maschler, Peleg, and Shapley [5]) and we give it here for completeness only.

The *imputation set* $\mathcal{I}(v)$ is the set of points $x \in \Re^N$ with $x_i \geq v(i)$ for all $i \in N$ and $(e_N, x) = v(N)$. If (N, v) is a cooperative game, the *kernel* $\mathcal{K}(v)$ of v is the set of points x in the imputation set $\mathcal{I}(v)$ for which if

$$\max\{v(S) - x(S) \mid i \in S \subset N \setminus \{j\}\} =: s_{ij}(x) \\ > s_{ji}(x) := \max\{v(S) - x(S) \mid j \in S \subset N \setminus \{i\}\},$$

then

$$x_j = v(j)$$

(Davis and Maschler [1]).

The following proposition shows that for a game (N, v) with nonempty core C the intersection of C with the kernel of (N, v) is only dependent on C and contains the nucleolus of (N, v).

Proposition 2.1. *If (N, v) is a cooperative game and C is the non-empty core of v then $\mathcal{K}(v) \cap C$ is determined by C only (and contains the nucleolus of v).*

In fact, the intersection $\mathcal{K}(v) \cap C$ consists of the points $x \in C$ for which

$$\max\{t \in \Re \mid x + te_i - te_j \in C\} + \min\{t \in \Re \mid x + te_i - te_j \in C\} = 0 \quad \forall i \neq j. \tag{2.1}$$

If we define the *kernel of the pre-core* C as the set given by condition 2.1 (notation: $\mathcal{K}(C)$), then we find that the kernel of C contains the nucleolus of every game (N, v) with core C, *i.e.*,

$$\mathcal{L}oc\mathcal{N}uc(C) \subset \mathcal{K}(C).$$

Before we can define the least core of a pre-core C, we have to introduce some terminology. Let C be a pre-core and (N, v_C) the exact game with core C. Then we can distinguish three kinds of coalitions: A coalition S is *tight* if $(e_S, x) = v_C(S)$ for all $x \in C$. A coalition S will be called *essential* if $C \cap \{x \mid (e_S, x) = v_C(S)\}$ has codimension 1 in the core C (determines a facet of C). Coalitions which are neither tight nor essential are called *redundant*. Notice that if C is full dimensional, only the grand coalition is tight and each facet is determined by exactly one essential coalition. This means that the values of redundant coalitions can be decreased without changing the core of the game. In the example of the first section the coalitions $S = N, \{12\}, \{23\}, \{34\}$, and $\{14\}$ are tight and, *e.g.*, the coalition $S = \{24\}$ is essential for the boundary point ("facet") $s = 1$. The exact game v_C gives to the coalitions (123) and (134) the value $\frac{5}{4}$ and to the coalitions (1) and (3) the value $\frac{1}{4}$. The other coalitions have the same values as in the example. Notice that in this example all coalitions are tight or essential.

Let C be a pre-core and $\mathcal{S}$ the family of coalitions which are tight or essential. Let $E: C \to \Re^{\mathcal{S}}$ be the map $x \to \{E_S(x)\}_{S \in \mathcal{S}}$ where $E_S(x) = v_C(S) - x(S)$. As usual in the definition of a nucleolus, $\theta: \Re^{\mathcal{S}} \to \Re^s$, $s := |\mathcal{S}|$ is the map that orders the coordinates in a weakly decreasing order and $\preceq_{\text{lex}}$ is the lexicographic ordering on $\Re^s$. The *least core* of C is the set of points in C where the pointwise maximum $\bigvee_{S \in \mathcal{S}} E_S$ of the excess functions $\{E_S\}_{S \in \mathcal{S}}$ takes its minimum value. The *nucleolus* of C are the points of C where $\theta \circ E$ takes its lexicographic minimum.

$$\mathcal{LC}(C) := \left\{ x \in C \mid \bigvee_{S \in \mathcal{S}} E_S(x) \leq \bigvee_{S \in \mathcal{S}} E_S(y) \text{ for all } y \in C \right\}$$

and

$$\mathcal{N}(C) := \{ x \in C \mid \theta \circ E(x) \preceq_{\text{lex}} \theta \circ E(y) \text{ for all } y \in C \} \ .$$

Since C is compact $\mathcal{LC}(C)$ and $\mathcal{N}(C)$ are non-empty. As the excess functions E_S, $S \in \mathcal{S}$ are constant on $\mathcal{N}(C)$ and the linear inequalities $(e_S, x) \geq a_S$ for tight or essential coalitions S determine the bounded set C, the nucleolus of C consists of one point (cf. Schmeidler [9] and Potters and Tijs [8]). Note that all the concepts needed to define the least core and the nucleolus of the pre-core C are uniquely determined by C itself.

In the second proposition we will prove that the least core of every game with core C (see Maschler, Peleg, and Shapley [5] and Maschler, Potters, and Tijs [6]) coincides with $\mathcal{LC}(C)$. As the nucleolus of a game is always a point of the least core of a game we find that also the set $\mathcal{LC}(C)$ is a subset of C which contains the locus of the nucleolus $\mathcal{L}oc\mathcal{N}uc(C)$.

Proposition 2.2. *The least core of each game (N, v) with core C equals $\mathcal{LC}(C)$ and is therefore not dependent on v.*

Proof: We may assume that C is full-dimensional. Otherwise $C = \mathcal{LC}(C)$ and there is nothing to prove. The least core of a game (N, v) consists of the optimal points x of the linear program

$$\begin{aligned} \text{Maximize } \ & t \\ \text{subject to } \ & t - x(S) \leq -v(S) \quad (\forall S \subset N) \\ & x(N) = v(N) \ . \end{aligned} \tag{2.2}$$

Let t be a real number in the segment $[0, \hat{t}]$ where $\hat{t}$ is the maximal value of t in the linear program (2.2) determining the least core. Let us consider the linear program

$$\begin{aligned} \text{Minimize } \ & x(T) \\ \text{subject to } \ & x(S) \geq v(S) + t \quad (\forall S \in \mathcal{S}) \\ & x(N) = v(N) \ , \end{aligned} \tag{2.3}$$

where T is any redundant coalition for C. We will prove that the minimal value of this program is at least $v(T) + t$. In that case the optimal points of the first program (2.2) do not change if we replace the constraints $t - x(S) \leq -v(S)$, $S \subset N$ by $t - x(S) \leq -v(S)$, $S \in \mathcal{S}$ *i.e.*, $\mathcal{LC}(v) =$

$\mathcal{LC}(C)$. The dual program of the linear program (2.3) is

$$\begin{aligned} &\text{Maximize } \sum_{S\in\mathcal{S}} y_S(v(S)+t)+zv(N) \\ &\text{subject to } y_S \geq 0,\ S\in\mathcal{S}, z\in\Re, \\ &\sum_{S\in\mathcal{S}} y_S e_S + z e_N = e_T\ . \end{aligned} \tag{2.4}$$

For $t = 0$, the maximal value of (2.4) (*i.e.*, the minimal value of (2.3)) is at least $v(T)$ (as feasible points of (2.3) are in $C = Core(v)$). If we take an optimal solution $(\hat{y}_S, S \in \mathcal{S}, \hat{z})$ for $t = 0$, then this point is also feasible for $t > 0$ and the value of the goal function is

$$\sum_{S\in\mathcal{S}} \hat{y}_S v(S) + \hat{z}v(N) + t\sum_{S\in\mathcal{S}} \hat{y}_S\ .$$

We are left to prove that $\sum_{S\in\mathcal{S}} y_S e_S + z e_N = e_T$ with $y_S \geq 0$, $S \in \mathcal{S}$, implies that $\sum_{S\in\mathcal{S}} y_S \geq 1$. By taking $j \notin T$, we find that $z \leq 0$ and if we take $i \in T$ we find

$$\sum_{S\in\mathcal{S}} y_S \geq \sum_{S\in\mathcal{S}:i\in S} y_S + z = 1.$$

∎

Summarizing the two preceding propositions we have

$$\mathcal{L}oc\mathcal{N}uc(C) \subset \mathcal{K}(C) \cap \mathcal{LC}(C)\ . \tag{2.5}$$

Notice that Example 1.1 shows that there is no equality in general.

Corollary 2.3. *If C is the core of a convex game, then*

$$\mathcal{L}oc\mathcal{N}uc(C) = \{\mathcal{N}(C)\}\ .$$

Proof: If C is the core of a convex game v, then $v_C = v$ is convex (Schmeidler [10]). For a convex game the kernel consists of one point, the nucleolus of the game (Maschler, Peleg, and Shapley [4]). This means that $\mathcal{K}(C)$ consists of one point and the same holds for the locus of the nucleolus by (2.5). Let (N, v) be an arbitrary balanced game and define a second game $(N, \tilde{v})$ by

$$\tilde{v}(S) := \max\{v(S), v(N) - M_v(N \setminus S)\}.$$

The vector M_v is the vector of marginals for the grand coalition, *i.e.*, $M_v(i) = v(N) - v(N \setminus \{i\})$ for all $i \in N$.

The games (N, v) and $(N, \tilde{v})$ have the same core.

This can be seen as follows. As $\tilde{v} \geq v$ and $\tilde{v}(N) = v(N)$, we find $Core(\tilde{v}) \subset Core(v)$. If, however, $x \in Core(v)$ we have $x(S) \geq v(S)$ for all $S \subset N$, and $x_i \leq M_v(i)$ for all $i \in N$. From the last inequalities follows that $x(S) = x(N) - x(N \setminus S) \geq v(N) - M_v(N \setminus S)$. Therefore, $x(S) \geq \tilde{v}(S)$ for every coalition $S \subset N$. ∎

If (N, v) is a balanced game and $(N, \tilde{v})$ is *convex* then the locus of the nucleolus of the core of (N, v) consists of one point, the nucleolus of (N, v). This follows from the corollary and the preceding remark. In the literature there are two classes of games satisfying the condition that $(N, \tilde{v})$ is convex namely the class of *k-convex games* (Driessen [2]) and the class of *clan games* (Potters, Poos, Tijs, and Muto [7]):

(1) k-convex games. Let k be a natural number $1 \leq k \leq n$. According to Driessen [2], a cooperative game (N, v) is *k-convex* if $v\, S) \leq v(N) - M_v(N \setminus S)$ for all coalitions S with $|S| \geq k$ and the game v_k defined by $v_k(S) = \tilde{v}(S)$ for $|S| \geq k$ and $v_k(S) = v(S)$ for $|S| < k$ is convex. If (N, v) is a k-convex game then $v \leq v_k \leq \tilde{v}$ and because v and $\tilde{v}$ have the same core, say C, also the convex game v_k has core C. Therefore, if C is the core of a k-convex game then C is also the core of a convex game. Consequently, the locus of the nucleolus consists of one point.

(2) Clan games. A game (N, v) is a *clan game* if $v \geq 0$, $M_v \geq 0$ and there is a nonempty coalition Cl (the clan) such that $v(S) = 0$ if $S \not\supset Cl$ (the clan condition) and $v(S) \leq v(N) - M_v(N \setminus S)$ if $S \supset Cl$ (the union condition) (Potters et al. [7]).

In a clan game the clan members have a strong incentive to stick together. Otherwise a positive result is not possible. The non-clan members have no possibility to obtain a positive result by their own but they can block a part of the profit to be made by the clan members and this blocking power is larger if they form a union before they start the negotiations with the clan.

For a clan game (N, v) with clan Cl the game $\tilde{v}$ has the values

$$\tilde{v}(S) = \begin{cases} 0 & \text{if } S \not\supset Cl \\ v(N) - M_v(N \setminus S) & \text{if } S \supset Cl \end{cases}.$$

The first line follows from $M_v \geq 0$ and $M_v(i) = v(N)$ if $i \in Cl$; the second line follows from the union condition. It is not difficult to see that $\tilde{v}$ is a convex game: For $S, T \subset N$ with $Cl \subset S \cap T$ we have $\tilde{v}(S) + \tilde{v}(T) = \tilde{v}(S \cup T) + \tilde{v}(S \cap T)$. If S or T doesn't contain the clan the convexity condition follows from the monotonicity of $\tilde{v}$. Hence, we can apply the corollary and find that the locus of the core of a clan game consists of one point too.

Remark 2.4. For k-convex games the game v_k is a convex game with the same core as (N, v). A convex game is exact and therefore, $v_k = v_C$, the unique exact game with core $C = Core(v)$. For clan games the game $\tilde{v}$ is convex and therefore $\tilde{v} = v_C$.

3. Some other properties of the locus of the nucleolus

Let C be a pre-core of full dimension (only the grand coalition is tight) and let $\mathcal{S}$ and $\mathcal{R}$ be the families of essential and redundant coalitions. The set of games (N, v) with core C can be described by the (in)equalities $v(S) = v_C(S)$ for $S \in \mathcal{S}$ and $v(R) \leq v_C(R)$ for $R \in \mathcal{R}$. Let $\Delta := \max_{S \in \mathcal{S}}(v_C(N) - v_C(S) - v_C(N \setminus S))$. In the next proposition we prove that $\mathcal{N}(C)$ is the nucleolus of (N, v) if $v(R) < v_C(R) - \Delta$ for all redundant coalitions $R \in \mathcal{R}$, *i.e.*, if the redundant coalitions have small values they have no longer influence on the position of the nucleolus.

Proposition 3.1. *If C is a pre-core of full dimension and (N, v) is a game with core C of which all redundant coalitions R have values smaller than $v_C(R) - \Delta$, then $\mathcal{N}(C) = Nu(v)$.*

Proof: The proof is based on two observations:
(1) If $v(R) < v_C(R) - \Delta$, then $v(R) - x(R) < v_C(S) - x(S)$ for all essential coalitions $S \in \mathcal{S}$ and all core elements $x \in C$.
(2) The positive cone generated by $\{e_S\}_{S \in \mathcal{S}}$ and $-e_N$ contains all points of $\Re^N$.

Proof of (1): Let $x \in C$ and $S \in \mathcal{S}$. Then

$$\Delta \geq v_C(N) - v_C(S) - v_C(N \setminus S) \geq x(S) - v_C(S)$$

because $v_C(N) = x(N)$ and $x(N \setminus S) \geq v_C(N \setminus S)$. For a redundant coalition $R \in \mathcal{R}$ we have

$$v(R) - x(R) < v_C(R) - x(R) - \Delta \leq -\Delta \leq v_C(S) - x(S).$$

Proof of (2): As C is compact and as it is determined by the linear relations

$$x(S) \geq v_C(S) \quad \text{for} \quad S \in \mathcal{S} \quad \text{and} \; x(N) = v_C(N)$$

each linear function $x \to (c, x)$ has a minimum value on C and the dual linear program is feasible for every vector $c \in \Re^N$. This means that $\sum_{S \in \mathcal{S}} y_S \, e_S + z \, e_N = c$ is solvable with $y_S \geq 0$ for $S \in \mathcal{S}$ and

$z \in \Re$. Therefore, every point of $\Re^N$ is a nonnegative linear combination of $\{e_S\}_{S \in \mathcal{S}}$ and $\{e_N, -e_N\}$. We can skip the vector e_N since the family $\mathcal{S}$ contains a balanced family (cf. Maschler, Potters, and Tijs [6]).

Let $x_0 \in C$ be the nucleolus of (N, v) and suppose that $x_0 \prec_{\text{lex}} \mathcal{N}(C)$. If we order the excesses of x_0 and $\mathcal{N}(C)$ in a weakly decreasing order, in both sequences the excesses of essential coalitions precede the excesses of the redundant coalitions (by observation (1)). From the definition of $\mathcal{N}(C)$ we infer that $x_0(S) = \mathcal{N}(C)(S)$ for all $S \in \mathcal{S}$. Then $x_0 = \mathcal{N}(C)$ by observation (2). ∎

Corollary 3.2. $\mathcal{N}(C) \in \mathcal{L}oc\mathcal{N}uc(C)$.

In the following proposition we give a necessary and sufficient condition that $\mathcal{L}oc\mathcal{N}uc(C)$ consists of the point $\mathcal{N}(C)$ only.

Let C be a full dimensional pre-core and be $\bar{v}_C$ the game with values $\bar{v}_C(S) = v_C(S)$ if $S = N$ or S is essential and $\bar{v}_C(R) = v_C(R) - \Delta$ if R is redundant. Let $x_0 = \mathcal{N}(C)$. Then x_0 is the nucleolus (and the prenucleolus) of $\bar{v}_C$. Let $\bar{\mathcal{B}}_t$ be the families of coalitions defined by

$$\{S \subset N \mid \bar{v}_C(S) - x_0(S) \geq t\}.$$

These families $\bar{\mathcal{B}}_t$ are balanced (as far as they are not empty) (Sobolev [11]). Let $[\bar{\mathcal{B}}_t]$ be the linear subspace of $\Re^N$ generated by $\{e_S \mid S \in \bar{\mathcal{B}}_t\}$. For each redundant coalition R, we define $t_R := v_C(R) - x_0(R)$. Using these notations we can formulate the following propsition:

Proposition 3.3. *If C is a pre-core of full dimension then $\mathcal{L}oc\mathcal{N}uc(C) = \{\mathcal{N}(C)\}$ if and only if $e_R \in [\bar{\mathcal{B}}_{t_R}]$ for every redundant coalition R.*

Proof: Let (N, v) be a game with core C and let $\mathcal{B}_t$ be the families of coalitions $\{S \subset N \mid v(S) - x_0(S) \geq t\}$. We prove that these families are balanced. For every redundant coalition $R \in \mathcal{B}_t$ we have $t_R \geq t$ (because $v_C(R) \geq v(R)$) and therefore, $\bar{\mathcal{B}}_{t_R} \subset \bar{\mathcal{B}}_t$ and $\mathcal{B}_{t_R} \subset \mathcal{B}_t$. Let $e_R = \sum_{S \in \bar{\mathcal{B}}_{t_R}} y_S^R e_S$ with $y_S^R \in \Re$. As $\bar{\mathcal{B}}_t$ is balanced we have $\sum_{T \in \bar{\mathcal{B}}_t} x_T e_T = e_N$ with positive coefficients x_T for all $T \in \bar{\mathcal{B}}_t$. Then

$$\sum_{T \in \bar{\mathcal{B}}_t} x_T e_T + \delta \sum_{R \in \mathcal{B}_t} e_R - \delta \sum_{R \in \mathcal{B}_t} \sum_{S \in \bar{\mathcal{B}}_{t_R}} y_S^R e_S = e_N.$$

For small but positive δ this gives the balancedness of $\mathcal{B}_t$. Then $\mathcal{N}(C) = x_0$ is the (pre)nucleolus of every game (N, v) with core C.

Conversely, if $e_R \notin [\bar{\mathcal{B}}_{t_R}]$ we construct a game (N, v) with core C and nucleolus $\neq \mathcal{N}(C)$. Let $v(S) = v_C(S)$ if $S = N, R$ or S is essential and

$v(S) = \bar{v}_C(S)$ if S is redundant $\neq R$. If $x_0 = \mathcal{N}(C)$ is the nucleolus of (N, v) then in particular $\mathcal{B}_{t_R} = \bar{\mathcal{B}}_{t_R} \cup \{R\}$ is balanced. The equality holds because $\bar{\mathcal{B}}_{t_R}$ does not contain *all* essential coalitions and therefore no redundant coalition $\neq R$ (cf. the proof of Proposition 3.1). This means that the equation $\sum_{S \in \bar{\mathcal{B}}_{t_R}} \hat{y}_S e_S = e_N$ as well as the equation $\sum_{S \in \bar{\mathcal{B}}_{t_R}} y_S e_S + z e_R = e_N$ have a solution with positive coefficients. This means that $z e_R \in [\bar{\mathcal{B}}_{t_R}]$ with $z > 0$. This is in contradiction with the assumption about e_R. ▮

Remark 3.4. This proposition is no longer true if C is not full dimensional. In the example of Section 1 every coalition is tight or essential. Therefore, one side of the equivalence is an empty condition but we have seen that the locus of the nucleolus consists of more than one point.

Finally, we prove that the locus is a path-connected subset of C.

Proposition 3.5. *For every pre-core C the locus of the nucleolus is path-connected.*

Proof: We prove that every point of the locus can be connected by a path to $Nu(v_C)$. Suppose that x is the nucleolus of a game (N, v) with core C. Then $v \leq v_C$ and all the games $v_t = t\,v + (1-t)\,v_C$, $t \in [0,1]$ have core C. As the nucleolus is a continuous function of the game v [Schmeidler (1969)] the path $t \in [0,1] \to Nu(v_t)$ is a (continuous) path connecting $Nu(v_C)$ with $Nu(v) = x$. ▮

Open question: Is the locus of the nucleolus always a convex subset of C?

4. Some instructive examples and final remarks

In this final section we give some examples answering some questions the reader may have. In the first example we show that the locus of the nucleolus may consist of one point also if the pre-core C is not the core of a convex game.

Example 4.1. Let C be the convex hull of the points $(2,0,2,0)$, $(1,1,2,0)$, $(0,2,1,1)$, $(1,1,0,2)$ and $(2,0,1,1)$. The values of v_C in the coalitions (124), (12), (14) and (1) are 2, 2, 1 and 0 respectively. Hence, the game v_C is not convex. Nevertheless it is easy to see that the point $(1,1,1,1)$ is the unique point in C where $s_{12} = s_{21}$ and $s_{14} = s_{41}$ with respect to v_C. This means that according to Proposition 2.1, the point $(1,1,1,1)$ is the only point in $\mathcal{K}(C) = \mathcal{L}oc\mathcal{N}uc(C)$.

The second example is a five person game which exhibits an important phenomenon which makes the locus problem so interesting but also so difficult.

Example 4.2. Let C be the core of the following game:

				18					
		L	L	L	L	9			
8	L	L	4	L	L	8	4	L	8
L	L	L	6	L	6	L	L	L	L
		L	L	L	L	L			

In each line the coalitions are ordered in lexicographic order and an "L" indicates that the value of this coalition is so low that it has no influence on the shape of the core and the position of the nucleolus. The facets of C are determined by the coalitions (123), (234), (345), (15), (24), (134) and (235). The coalition (2345) is redundant. It is not difficult to see that the least core of this game is the set

$$\{(t+2, 6-t, 2, t+2, 6-t) \,|\, 0 \leq t \leq 4\} .$$

The end points of the least core are determined by the coalitions (134) for $t = 0$ and (235) for $t = 4$. The hyperplane $x(2345) = 11$ has no points in common with the least core (cf. the proof of Proposition 2.1). The excess of coalition (2345) has decreased faster than the excesses of the other coalitions. The excesses of the coalitions (123), (234), (345), (15) and (24) are constant on the least core. Therefore, the position of the nucleolus in the least core is determined by the coalitions (134), (235) and (2345) and one might expect that the last coalition has no influence. But in fact the excesses of those three coalitions are $-2-2t$, $2t-10$ and $-7+t$ respectively. The nucleolus corresponds to $t = \frac{5}{3}$ and coalition (235) has a smaller excess in that point than (2345). So in the shrinking process first the excess of (2345) decreases faster than the excess of (235) but after reaching the least core the excess of (2345) decreases slower than the excess of (235). This phenomenon finds its origin in the equality

$$e_{\{2345\}} = \tfrac{1}{2}e_{\{235\}} + \tfrac{1}{2}e_{\{24\}} + \tfrac{1}{2}e_{\{345\}}.$$

If the payoffs $x(S)$ to the coalitions $S =$ (235), (24) and (345) increase with δ the payoff to coalition (2345) increases with $\frac{3}{2}\delta$ (faster) but after reaching the least core the payoffs to (24) and (345) do not change anymore. If the payoff to (235) increases with δ the payoff to coalition (2345) increases with $\frac{1}{2}\delta$ (slower).

Summary: In this paper we investigated the locus of the nucleolus in a pre-core C. We proved that this locus consists of one point if the set C is